Lecture Notes
in Business Information Processing

384

More information about this series at http://www.springer.com/series/7911

José María Moreno-Jiménez ·
Isabelle Linden · Fatima Dargam ·
Uchitha Jayawickrama (Eds.)

Decision Support Systems X

Cognitive Decision Support Systems and Technologies

6th International Conference on Decision
Support System Technology, ICDSST 2020
Zaragoza, Spain, May 27–29, 2020
Proceedings

 Springer

Editors
José María Moreno-Jiménez ⓘ
University of Zaragoza
Zaragoza, Spain

Isabelle Linden ⓘ
University of Namur
Namur, Belgium

Fatima Dargam ⓘ
SimTech Simulation Technology
Graz, Austria

Uchitha Jayawickrama ⓘ
Loughborough University
Loughborough, UK

ISSN 1865-1348   ISSN 1865-1356 (electronic)
Lecture Notes in Business Information Processing
ISBN 978-3-030-46223-9   ISBN 978-3-030-46224-6 (eBook)
https://doi.org/10.1007/978-3-030-46224-6

This Springer imprint is published by the registered company Springer Nature Switzerland AG
The registered company address is: Gewerbestrasse 11, 6330 Cham, Switzerland

EURO Working Group on Decision Support Systems

The EWG-DSS is a Euro Working Group on Decision Support Systems within EURO, the Association of the European Operational Research Societies. The main purpose of the EWG-DSS is to establish a platform for encouraging state-of-the-art high-quality research and collaboration work within the DSS community. Other aims of the EWG- DSS are to:

- Encourage the exchange of information among practitioners, end-users, and researchers in the area of decision systems
- Enforce the networking among the DSS communities available and facilitate activities that are essential for the start up of international cooperation research and projects
- Facilitate the creation of professional, academic, and industrial opportunities for its members
- Favor the development of innovative models, methods, and tools in the field of decision support and related areas
- Actively promote the interest on decision systems in the scientific community by organizing dedicated workshops, seminars, mini-conferences, and conference, as well as editing special and contributed issues in relevant scientific journals

The EWG-DSS was founded with 24 members, during the EURO Summer Institute on DSS that took place at Madeira, Portugal, in May 1989, organized by two well-known academics of the OR community: Jean-Pierre Brans and José Paixão. The EWG-DSS group has substantially grown along the years. Currently, we have over 300 registered members from around the world.

Through the years, much collaboration among the group members has generated valuable contributions to the DSS field, which resulted in many journal publications. Since its creation, the EWG-DSS has held annual meetings in various European countries, and has taken active part in the EURO Conferences on decision-making-related subjects. Starting from 2015, the EWG-DSS established its own annual conferences, namely, the International Conference on Decision Support System Technology (ICDSST).

The current EWG-DSS Coordination Board comprises eight experienced scholars and practitioners in DSS field: Pascale Zaraté (France), Fátima Dargam (Austria), Shaofeng Liu (UK), Boris Delibašić (Serbia), Isabelle Linden (Belgium), Jason Papathanasiou (Greece), Pavlos Delias (Greece) and Ana Paula Cabral Seixas Costa (Brasil).

Preface

This 10th edition of the EWG-DSS Decision Support Systems published in the Springer LNBIP series presents the full papers from the 6th International Conference on Decision Support System Technology (ICDSST 2020) held in Zaragoza, Spain, during May 27–29, 2020, with the main topic "Cognitive Decision Support Systems & Technologies." This event was organized by the Euro Working Group on Decision Support Systems (EWG-DSS) in collaboration with the University of Zaragoza. Exceptionally, following the coronavirus crisis, the 2020 edition of the conference was held in an online mode.

This conference series was starting with ICDSST 2015 in Belgrade, where it was planned to consolidate the tradition of annual events organized by the EWG-DSS in offering a platform for European and international DSS communities, comprising the academic and industrial sectors, to present state-of-the-art DSS research and developments, to discuss current challenges that surround decision-making processes, to exchange ideas about realistic and innovative solutions, and to co-develop potential business opportunities.

The scientific topical areas of ICDSST 2020 include:

- Decision Support Systems: Advances and Future Trends
- Multi-Attribute and Multi-Criteria Decision Making
- Knowledge Acquisition, Management, Extraction, Visualization, and Decision Making
- Multi-Actor Decision Making: Group and Negotiated Decision Making
- Collaborative Decision Making and Decision Tools
- Discursive and Collaborative Decision Support Systems
- Mobile and Cloud Decision Support Systems
- Applied Decision Support Systems
- GIS and Spatial Decision Support Systems
- Data Science, Data Mining, Text Mining, and Sentimental Analysis
- Big Data Analytics
- Imaging Science (Image Processing, Computer Vision, and Pattern Recognition)
- Human-Computer Interaction
- Internet of Things
- Social Network Analysis for Decision Making
- Simulation Models and Systems, Regional Planning, and Logistics and Traceability
- Smart City Mobility
- Business Intelligence and Quantum Economy
- Machine Learning, Natural Language Processing, and Artificial Intelligence
- Virtual and Augmented Reality
- New Methods and Technologies for Cognitive Decision Making
- Cognitive DSS and Cognitive Technologies
- Affective DSS and Affective Technologies

– Cognitive Computing
– Cognitive Applications (Cognitive Democracy and Participation, Cognitive Risk Management, etc.)
– General DSS Case Studies (Business Intelligence, Education, E-Government, Energy, Entrepreneurship, Environment, Health Care, Industrial Diversification and Sustainability, Innovation, Logistics, Natural Resources etc.)

This wide and rich variety of themes allowed us, in the first place, to present a summary about some solutions regarding the implementation of decision-making process in a high variety of domains and, in the second place, to highlight their main trends and research evolution. Moreover, this EWG-DSS LNBIP Springer edition has considered contributions selected from a triple-blind paper evaluation method, maintaining this way its traditional high-quality profile. Each selected paper was reviewed by at least three internationally known experts from the ICDSST 2020 Program Committee and invited external reviewers. Therefore, through its rigorous two-stage based triple-round review, 13 out of 51 submissions, which corresponds to a 25.5% acceptance rate, were selected in order to be considered for inclusion this 10th EWG-DSS Springer LNBIP Edition.

In this context, the selected papers are representative of the current and relevant DSS research and application advances. The papers are organized in three sections:

(1) Methodological Contributions. Five papers are presented in this section: "Two Sides of Collective Decision Making - Votes from Crowd and Knowledge from Experts" by Zorica A. Dodevska, Ana Kovacevic, Milan Vukicevic, and Boris Delibašić. "Using Multi-level DEA to go beyond the three dimensions of sustainability" by Georgios Tsaples and Jason Papathanasiou. "Analysis of Graphical Visualizations for Multi-Criteria Decision Making in FITradeoff Method Using a Decision Neuroscience Experiment" by Lucia Reis Peixoto Roselli and Adiel Teixeira de Almeida. "The Multi-Actor Multi-Criteria Analysis (MAMCA): New Software and New Visualizations" by He Huang, Philippe Lebeau, and Cathy Macharis. And lastly, "Complexity Clustering of BPMN Models: Initial Experiments with the K-means Algorithm" by Chrysa Fotoglou, George Tsakalidis, Kostas Vergidis, and Alexander Chatzigeorgiou.

(2) Case Studies. This section opens with a case study application in Private Digital Security: "To Click or Not to Click? Deciding to Trust or Distrust Phishing Emails" by Pierre-Emmanuel Arduin. Then, three case studies from the Public Sector are presented: "Management Information System for Police Facility Location" by Ana Paula Henriques de Gusmão, Bruno Ferreira da Costa Borba, and Thárcylla Rebecca Negreiros Clemente; "An Automated Corpus Annotation Experiment in Brazilian Portuguese for Sentiment Analysis in Public Security" by Victor Diogho Heuer de Carvalho, Thyago Celso Cavalcante Nepomuceno, and Ana Paula Cabral Seixas Costa; and "An E-government Procurement Decision Support System Model for Public Private Partnership Projects in Egypt" by Karim Soliman and Nada El-Barkouky. Finally, there are three cases studies from the Private Sector: "Determinants of recommendation in the airline industry: An application of online review analysis" by Praowpan Tansitpong; "Artificial Intelligence Based Decision Making for Venture Capital Platform" by Ankit

Tewari, Joaquim Gabarro, Josep Sole, Brice Lapouble, and Lluis Montull; and "Supporting Operational Decisions on Desalination Plants from Process Modelling and Simulation to Monitoring and Automated Control with Machine Learning" by Fatima Dargam, Erhard Perz, Stefan Bergmann, Ekaterina Rodionova, Pedro Sousa, Francisco Alexandre A. Souza, Tiago Matias, Juan Manuel Ortiz, Abraham Esteve-Nuñez, Pau Rodenas, and Patricia Zamora Bonachela.

(3) The last section involves a state-of-the-art overview paper: "DSS, BI, and Data Analytics Research: Current State and Emerging Trends (2015–2019)"

We would like to thank many people, who greatly helped to achieve the success of this LNBIP book. First of all, we would like to thank Springer for giving us the opportunity to guest edit the DSS book, and we especially wish to express our sincere gratitude to Alfred Hofmann and Anna Kramer, who have provided us with timely professional guidance and advice during the volume editing process. Secondly, we need to thank all the authors for submitting their state-of-the-art work to be considered for the LNBIP volume. All selected papers are of extremely high quality. It was a hard decision for the guest editors to select the best 13. Thirdly, we wish to express our gratitude to the reviewers, who volunteered to help with the selection and improvement of the papers.

Finally, we believe that this EWG-DSS Springer LNBIP volume has made a high-quality contribution of well-balanced and interesting research papers addressing the conference's main theme. We hope the readers will enjoy the publication!

May 2020

José María Moreno-Jiménez
Isabelle Linden
Fatima Dargam
Uchitha Jayawickrama

Organization

Program Committee

Carlos Antunes	INESC Coimbra, University of Coimbra, Portugal
Francisco Antunes	INESC Coimbra and Beira Interior University, Portugal
Dragana Becejski-Vujaklija	University of Belgrade, Serbia
Marko Bohanec	Jožef Stefan Institute, Slovenia
Guy Camilleri	IRIT IC3, Université Paul Sabatier, France
Jesús Cardeñosa	Universidad Politécnica de Madrid, Spain
Hing Kai Chan	University of Nottingham, UK
Alok Choudhary	Loughborough University, UK
Christian Colot	University of Namur, Belgium
Fatima Dargam	SimTech Simulation Technology, ILTC, Austria
Adiel Teixeira de Almeida	Federal University of Pernambuco, Brazil
Pavlos Delias	International Hellenic University, Greece
Boris Delibašić	University of Belgrade, Serbia
Panagiota Digkoglou	University of Macedonia, Greece
Alex Duffy	University of Strathclyde, UK
Sean Eom	Southeast Missouri State University, USA
María Teresa Escobar	Universidad de Zaragoza, Spain
Carlos Flavian	Universidad de Zaragoza, Spain
Gabi Florescu	ICI, Romania
Jorge Freire De Sousa	University of Porto, Portugal
Jean-Marie Jacquet	University of Namur, Belgium
Uchitha Jayawickrama	Loughborough University, UK
Daouda Kamissoko	EMAC, France
Marc Kilgour	Wilfrid Laurier University, Canada
Kathrin Kirchner	Technical University of Denmark, Denmark
Emilio Larrodé	Universidad de Zaragoza, Spain
Isabelle Linden	University of Namur, Belgium
Shaofeng Liu	University of Plymouth, UK
João Lourenço	Universidade de Lisboa, Portugal
Jan Mares	UCT Prague, Czech Republic
Bertrand Mareschal	ULB, Belgium
Nikolaos Matsatsinis	Technical University of Crete, Greece
José María Moreno-Jiménez	Universidad de Zaragoza, Spain
Jason Papathanasiou	University of Macedonia, Greece
Gloria Phillips-Wren	Loyola University Maryland, USA
Francois Pinet	Cemagref, France
Nikolaos Ploskas	University of Western Macedonia, Greece
Sandro Radovanović	University of Belgrade, Serbia

Contents

Overview

Methodological Contributions

Two Sides of Collective Decision Making - Votes from Crowd and Knowledge from Experts

Zorica A. Dodevska$^{(\boxtimes)}$ (iD), Ana Kovacevic, Milan Vukicevic (iD), and Boris Delibašić (iD)

Faculty of Organizational Sciences, University of Belgrade,
154 Jove Ilića, 11000 Belgrade, Serbia
{zd20185004,ak20195017}@student.fon.bg.ac.rs,
{milan.vukicevic,boris.delibasic}@fon.bg.ac.rs

Abstract. This paper deals with the role of experts and crowds in solving important societal issues. The authors argue that both experts and crowds are important stakeholders in collective decision making which should jointly participate in the decision-making process to improve it. Usually studied in different research areas, there have been a few models that integrate crowds and experts in a joint model. The authors give an overview of the advantages and disadvantages of crowd and expert decision making and highlight possibilities to connect these two worlds. They position the research in the area of Computational Social Choice (COMSOC) and crowd voting, emerging fields that bring great potential for collective decision making. COMSOC focuses on improving social welfare and the quality of products and services through the inclusion of community or clients into the decision-making process. Despite these altruistic goals, there are several shortcomings that call for the engagement of experts in voting procedures. The authors propose a simple participatory model for weighting and selection of voters and votes through the integration of expert rankings into crowd voting systems.

Keywords: Crowd · Experts · Collective decision making · Voting · Computational Social Choice (COMSOC) · Participatory models

1 Introduction

Voting is an important instrument for collective decision making, besides argumentation, fair allocation, and judgment aggregation [1]. Although often used to address important societal issues, various aspects of its imperfection are described in the literature and still need to be overcome – how to satisfy desirable properties of voting rules (listed in [2]), how to prevent strategic voting, manipulation, and bribery [3–5], how to design efficient algorithms for computationally NP-hard problems [6], or how to address preferential dependencies [7], are just some of questions. In order to understand contemporary trends of making collective decisions, the aim of this paper is to summarize two sides of contributions to the topic – votes that come from the general population or crowd, and knowledge from experts.

Participatory models of democracy, the huge impact of the Internet on society and technology (e.g., existence of online platforms, social media, Web 2.0, Big data), the

© Springer Nature Switzerland AG 2020
J. M. Moreno-Jiménez et al. (Eds.): ICDSST 2020, LNBIP 384, pp. 3–14, 2020.
https://doi.org/10.1007/978-3-030-46224-6_1

present trend of globalization (accompanied by decentralization, global collaboration, crowdsourcing) enable voting from the crowd – "crowd" in the sense of a large number of individuals who participate in order to make collective decisions. Hence, the role of "ordinary" voters is becoming increasingly prominent. Although experts traditionally have an important role in decision making, there is suspicion whether their role is getting increasingly marginalized or it becomes obvious that experts are even more necessary.

The paper has two directions/goals:

- The first goal is to summarize the literature about:

  - Different roles of experts and the crowd for making collective decisions;
  - The importance of a promising research field known as Computational Social Choice (COMSOC) for collective decision making;
  - Importance of public opinion and shortcomings of crowd voting;
  - The need for involving experts in voting procedures.

- The second goal is to propose a model for future research, based on the studied literature framework on the given topic, that integrates the advantages of expert and crowd voters.

2 Different Roles of Experts and the Crowd for Making Collective Decisions

Collective decision making refers to the process of making joint decisions, such that as many people as possible be satisfied with the reached outcome. It is most often achieved by collecting votes that represent voters' preferences regarding the offered alternatives (candidates, proposals, ideas, solutions, etc.) and by aggregating them.

In prominent voting rules (according to [8]): Borda count – counting the number of times a candidate ranked lower, Condorcet method – counting pairwise victories; voters give preference order to given alternatives, whereby the following conditions need to be satisfied [9]: total order of the alternatives, antisymmetric property, transitive property. Collective decisions are also often achieved through dialogue and consensus, where actors or agents take an active part in decision making.

Comparing crowd voting and expert ranking, Chen et al. [10] conclude that there are big differences regarding "who" gives votes – in the first case they are numerous voters that may have a very different educational background, while experts are small groups with specialized knowledge. The same authors point out that one may be a voter/selector and a candidate/contestant from the same crowdsourcing community, while experts are often recruited outside of it. Finally, the authors notice that there are differences in the way how the voting process is conducted: while experts are invited to follow an established process and given evaluation criteria, crowd voting is an open and dynamic process (in the sense that voters' criteria and the number of voters change over time).

Experts are individuals with developed intuition and reasoning for complex problem solving [11]. The crowd exchange opinions, experiences, and knowledge on the web and can contribute to collective decisions by synergy effects of a digital collaboration known in the literature as collective intelligence and "wisdom of the crowd" [12].

Expert voting is traditionally entrusted for solving complex problems in decision making or problems when professional knowledge from a particular field (e.g., technical, ethical, etc.) is required. Well-known decision-making methods, for example, multi-criteria decision-making (MCDM) methods require experts for the weighting of criteria and ranking of alternatives. MCDM methods can have complex procedures in situations when a multitude of criteria and sub-criteria with mutual interconnections are involved [13]. Crowd voting is rather applicable to common topics known to the wider population, on which every voter can have its own opinion expressed as a preference, which is a central topic in Artificial Intelligence (AI) [14].

There are challenges from both sides: experts may have difficulties to reach consensus, especially in the case of multidisciplinary decision making where availability of expertise is not a guarantee of its use [15]; while crowd voting can suffer from an inability to satisfy all desirable voting properties [16, 17].

Table 1. Experts vs. crowd – different aspects of collective decision making

Different aspects		Experts	Crowd
Voting properties [10]:	Selector qualification	Small groups of qualified judges	Many voters with unknown qualifications
	"Selector-selectee" relationship	Experts as an independent body	Voters and candidates can be from the same crowdsourcing community
	Selection process	A systematic voting process with justified evaluation criteria	An uncertain voting process in which participants have their own evaluation criteria
Decision making relies on...		...intuition and reasoning at the same time [11]	...collective intelligence and "wisdom of the crowd" [12]
Application		Complex decision-making problems (e.g., MCDM methods with mutual interdependencies of criteria [13])	Social choice topics with respect to individual preferences as a central topic of AI [14]
Challenges		Possible difficulties in multidisciplinary decision making [15]	Impossibility theorems in Social Choice Theory (e.g., Arrow's theorem [16], the Condorcet paradox [17])

Different aspects of collective decision making between experts and the crowd are summarized in Table 1.

3 The Importance of COMSOC for Collective Decision Making

Because collective decision making "takes place in the dual socio-technical environment, where humans and algorithms parallelly make decisions substantial for society" [18], both expert role and crowd role hinge together in COMSOC. As an interdisciplinary field extending from Social Choice Theory to Computer Science [19], and therefore, specific research area, COMSOC connects several fields that can be organized into the following heterogeneous layers:

1. Social Choice Theory and related social disciplines – Economics, Political Science, Philosophy, Psychology. These studies are focused on preferential listening and interested in achieving justice and social welfare.
2. Computer Science and related disciplines – AI with Machine learning, Mathematics, Data Science, Operations research. Some of the major aspects in this research direction are associated with the aggregation of preferences with the help of advanced algorithms, optimization and learning from data.

Similarly, as we can make an analogy between humans and intelligent agents, the first layer, i.e. Social Choice Theory, is a foundation of studying autonomous agents in multi-agent systems [20], and therefore it has found rich applications in the second layer of COMSOC. Assumed double-dealing of agents on the relation of its own and collective well-being are possible topics that the multi-agent system approach brings.

COMSOC manifesto from Conte et al. [21] emphasizes the role of COMSOC in addressing important issues in society. Some of them that they list are demographic growth, global crises, health threats, unethical behavior, misuse of information and privacy, which are the subject of collective decision making. In addition, COMSOC offers a new perspective on social choice problem solving, it stimulates the development of new computational methods based on traditional social choice concepts [22] (e.g., cutting-edge ensemble-based algorithms in Machine learning).

4 Importance of Public Opinion and Shortcomings of Crowd Voting

4.1 About the Importance of Public Opinion and Crowd Voting

Gaining insight into public opinion is motivated by the improvement of public satisfaction on common issues and achieving social welfare. The collection of votes from the general population can be encouraged by the following main reasons:

- Democratic participating in political elections and policymaking (e.g., law regulation [23]);

- Solving issues from common interest (e.g., budget allocation – Knapsack voting and participatory budgeting [24], or resolving a different kind of issues in the field of education, health, ecology, human rights, urban planning, etc.).

On the other side, companies see a great opportunity in crowd voting to increase the quality of products and services (and thus customer satisfaction) and reduce costs simultaneously. This is achieved by collecting and aggregating customer opinions, ideas and comments and including them into product/service development. Even more, nowadays companies often transfer part of the job to their users. For example, software companies publish their software very early in the testing phase and make updates based on user logs and comments (saving money and time for their testers and developers). In general, there is a number of cases where companies are interested in customer vote (opinion) collection and aggregation. Some of the interesting ones are the follows:

- Choosing innovative ideas or innovative solution that should be adopted, public supporting of innovative activities (open innovation and co-creation [25]);
- Giving feedback on creative works. Chen et al. [10] list some crowdsourcing platforms that use crowd voting in the field of design: *Zooppa* platform for marketing creatives (https://www.zooppa.com/en/), *Lego Ideas* website intended to commercialize the best users' ideas (https://ideas.lego.com/), *Jovoto* co-creation platform for brand innovation (https://www.jovoto.com/);
- Making recommendations based on users' critical rating (e.g., Amazon.com for e-commerce recommendation [26]);
- Selecting winners in competitions (e.g., TV music competitions such as Eurovision Song Contest, American Idol, etc.).

Therefore, it seems that crowd voting, as an alternative in crowdsourcing [27], when the crowd is asked to give subjective opinions about given alternatives in order to aggregate their votes, is an important instrument on the way for creating public opinion and a sound basis for making collective decisions.

4.2 Shortcomings of Crowd Voting

Despite a lot of advantages of crowd voting, that in many cases can successfully meet the needs of public opinion, there are still a lot of challenges that need to be surpassed.

The basic challenge for the successful exploitation of crowd voting is efficient collecting of votes. Cost assessment needs to be done. It is necessary to estimate the costs of usage and development of the COMSOC system, as well as the benefits (social or financial) that can be achieved. The good news is that there is a lot of free platforms that can be exploited, but also platforms that can collect votes at reasonable prices.

One noticeable problem is an unknown level of competence of ordinary voters. Unlike expert voters who have profound knowledge about the subject of voting, public voters may have insufficient information or knowledge regarding a given subject. The level of competence of ordinary voters is usually not verified, which is related to the problem of the quality of the voting outcome.

Another problem that is characteristic of crowdsourcing, in general, is the "digital divide" [28]. This means that only those who have access to the Internet can participate in online voting. In that sense, online crowd voting could not be considered as a fully representative process of democracy.

Even without the digital divide considered, crowed voting can suffer from a collection of a non-representative sample of votes, which may be the result of the following problems:

- There are strategies for single-agent manipulation and group manipulation in judgment aggregation [29].
- Bribery is recognized as an important problem in COMSOC [30] where financial incentives bring corrupt votes.
- Favoritism and bias regarding the choice of preferences are the consequence of subjective judgment, prejudice (toward individuals) and human tendencies to build stereotypes (toward groups of people).
- It is necessary to find people who want to participate. Potential voters do not want to participate, or they are not informed about voting. Additionally, voting turnout may be induced by complicated voting procedures (e.g., registration), multi-stage voting processes or lack of motivation to participate.

Quality Assessment and Control in Crowd Voting (An Analogy with Crowdsourcing). The quality of crowd voters (crowdsourced data and solutions in general) is highly susceptible to many factors and thus it is utterly important to address this problem adequately. Quality estimation and control of crowdsourcing based solutions is a very challenging task and many potential solutions are recently proposed [31–34]. Extensive survey [35] provides a comprehensive categorization of quality assessment techniques:

- **Individual.** In this approach, individuals are involved in the evaluation of co-workers and/or their results for specific tasks. In this case, individuals that are involved in assessment may be workers themselves, experts or requesters. Individual quality assessments may be created through [35]: ratings [33], qualification tests [34], self-correction [36], personality tests [37, 38], referrals [39, 40], expert reviews [41].
- **Group.** Group assessment methods may be equalized with voting as a basis for assessment. This means that groups may asses the quality of voters or other crowdsources (e.g., workers on labeling tasks.).
- **Computation-based.** This approach is trying to provide methods that will automate the quality of crowdsourced data through computation and machine learning methods. These methods will be addressed in this paper through the proposal of automated weighing and validation of votes in crowd voting.

Even though computation-based methods are classified as a separate category, it is important to note that both individual and group-based quality assessment techniques are in most cases powered by computational methods. For example, any group-based technique requires aggregation of group assessments and thus computational methods. This is especially important since evaluation assessments may drastically increase the cost of crowdsourcing (e.g., if a large group of workers or inclusion of expert opinions is integrated into the evaluation process). In many cases, this cost increase may reduce the basic advantages of COMSOC [32].

5   The Need for Involving Experts in Crowd Voting Procedures

In an ideal case, all crowd voters could be experts in the wanted area and they would be unbiased and motivated for voting. This means that voters would just express their pure preferences and the problem of COMSOC would be reduced only to vote aggregation.

One of the major problems of involving experts is their cost. On the other side, involving experts in voting procedures would largely benefit the quality of collective decision making.

Engaging participants for generating ideas from the crowd is a common problem, but it is shown that if experts provide a few alternative ideas then participants are more involved and generate rational ideas [42]. Also, the hybrid crowd (led by experts) shows advantages compared to the pure crowd [43].

Context respect when choosing a voting rule is an important question. Experts should analyze both the nature of social choice problematic and the computational pros and cons of a given voting rule.

Designing algorithms for aggregating votes is a technically demanding task due to complexity and therefore requires expert engagement. Multi-winner voting (an election of a set of winners), as one of the highlighted research areas in COMSOC [44], is a kind of complex voting. Many committee scoring rules used in this research direction are NP-hard in the sense of their computation [45]. On the other hand, computational complexity can be useful for finding NP-hard voting rule which works against manipulation, as an obstacle for cheating [46].

Besides the necessity of experts in crowd voting, this process clearly benefits if experts could weigh or exclude voters that are inconsistent or provide rankings that would lead to non-productive solutions. Of course, the involvement of experts in this process for all voters is not realistic since crowd voting assumes many voters. But, using limited expert input in synergy with computational methods could potentially increase the quality of crowdsourced data (in this case votes) as well as their aggregation.

Real-world applications that address both preferences from crowd and experts assembled in participatory models have a different domain of applications, and some examples include the election of software configuration (see [47]), *e-cognocracy* for new electronic democracy (see [48]), choosing winner in Eurovision Song musician contest (see [49]), deciding which art project to fund (see [50]), computational urban planning (see [51]), *MicroTalk* argumentation with 'true/false' answers (similar as approval voting) for improving accuracy of crowd workers (see [52]).

6   A Proposal for Future Research

The authors propose a simple participatory model for the integration of expert knowledge and ranking problems in crowd voting (see Fig. 1). The idea is to use the best from two different worlds. On the one hand, there are experts who give their ranks in choosing alternatives. On the other hand, there is a crowd that wants to participate in decision making. In the proposed model, the first step is filtering of crowd participants based on their ranking consistency (upper part of Fig. 1). Participatory voters should provide direct rankings as well as pairwise preferences for available alternatives. Consistency of voting may be assessed by rank comparison metrics (e.g., Spearman's rank correlation coefficient). Inconsistent participants are then filtered out from the voting process. Consistent crowd votes and expert rankings can be aggregated in the following way (middle part of Fig. 1): preference graphs are built from experts and crowd rankings. Nodes of the graph represent alternatives, while edges represent preference weights that are derived from rankings (e.g., the average number of voters) or explicit weights (that could be provided from experts). Experts' constraints may be incorporated by assigning infinite value to some edges (e.g., experts demand that alternative A is better ranked than alternative B). After the definition of expert and knowledge graphs, these graphs are fused in a common graph (e.g., by multiplication of expert and crowd edge weights). In this phase, experts are able to inspect fused graphs and reconsider their rankings (reweight expert graph) if the crowd is consistent with some rankings that differ from their initial opinions. In this way, experts can enable the creation of a solution that is

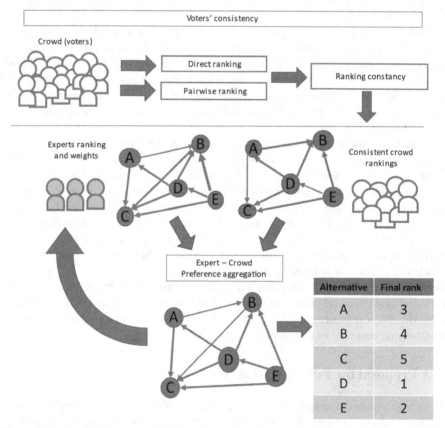

Fig. 1. A simple participatory model for the integration of expert knowledge and ranking problems in crowd voting

acceptable to them, while satisfying a high percentage of crowd participants. Finally, based on the fused graph final rankings are derived. Derivation of final rankings from the graph may be done by calculation of eigenvalues of graph incidence matrix similar to the PageRank algorithm or by conditional preference (CP) aggregation methods (such as CP-nets). Some applications of the PageRank algorithm are used for collaborative ranking frameworks suggested in [53, 54]. The proposed model in this paper includes such adjustment that tends to satisfy most voters' preferences while at the same time relies on experientially learned complex preferential dependencies validated by experts.

7 Conclusion and Future Work

Having in mind the different roles of experts-based and crowd-based decision making, this paper summarizes literature about making collective decisions from the perspectives of COMSOC and crowd voting. Even though crowd voting enables a number of benefits (both financial and social), it still suffers from many vulnerabilities like the engagement of participants, quality and consistency of votes (including the digital divide, bias, bribery, etc.). Identification and resolution of such problems is a very hard task and, in many cases, impossible with computational methods and COMSOC. On the other side, including expert knowledge, at the level that is achievable considering costs, could potentially improve enhance computational methods and quality of COMSOC and crowd-voting solutions.

In this paper, the authors proposed an idea of exploiting expert knowledge for weighting and selection of votes and voters as well for the estimation of voters' consistency. In future work, the authors plan to elaborate on this idea and provide an experimental evaluation of the system. They plan to validate the model in COMSOC problems of optimal team composition and participatory budgeting. The approach could be used for solving various real-life problems, one being improving the curriculum of an academic course (the idea is to use both students' preferences about a wide range of topics from the course area and academics' knowledge and experience, for example, about past students' acceptance of specific topics, compliance of program, or prerequisite knowledge on a specific level of studies).

Acknowledgements. This paper is a result of the project ONR - N62909-19-1-2008 supported by the Office for Naval Research, the United States: *Aggregating computational algorithms and human decision-making preferences in multi-agent settings.*

References

1. Baumeister, D., Rothe, J., Selker, A-K.: Strategic behavior in judgment aggregation. In: Endriss, U. (ed.) Trends in Computational Social Choice, pp. 145–168. AI Access (2017)
2. Cornelio, C., Pini, M.S., Rossi, F., Venable, K.B.: Multi-agent soft constraint aggregation via sequential voting: theoretical and experimental results. Auton. Agent. Multi-Agent Syst. **33**, 159–191 (2019)
3. Endriss, U., Obraztsova, S., Polukarov, M., Rosenschein, J.S.: Strategic voting with incomplete information. In: Proceedings of the Twenty-Fifth International Joint Conference on Artificial Intelligence (IJCAI 2016), pp. 236–242 (2016)

4. Faliszewski, P., Reisch, Y., Rothe, J., Schend, L.: Complexity of manipulation, bribery, and campaign management in Bucklin and fallback voting. Auton. Agent. Multi-Agent Syst. **29**, 1091–1124 (2015)
5. Faliszewski, P., Rothe, J.: Control and bribery in voting. In: Brandt, F., Conitzer, V., Endriss, U., Lang, J., Procaccia, A.D. (eds.) Handbook of Computational Social Choice, pp. 145–168. Cambridge University Press, New York (2016)
6. Skowron, P., Yu, L., Faliszewski, P., Elkind, E.: The complexity of fully proportional representation for single-crossing electorates. Theoret. Comput. Sci. **569**, 43–57 (2015)
7. Airiau, S., Endriss, U., Grandi, U., Porello, D., Uckelman, J.: Aggregating dependency graphs into voting agendas in multi-issue elections. In: Walsh, T. (ed.), IJCAI-11: Proceedings of the Twenty-Second International Joint Conference on Artificial Intelligence: Barcelona, Catalonia, Spain, 16–22 July 2011, vol. 1, pp. 18–23.: AAAI Press/International Joint Conferences on Artificial Intelligence, Menlo Park (2011)
8. Lee, D.T., Goel, A., Aitamurto, T., Landemore, H.: Crowdsourcing for participatory democracies: efficient elicitation of social choice functions. In: Proceedings of the Second AAAI Conference on Human Computation and Crowdsourcing (HCOMP 2014), pp. 133–142 (2014)
9. Slavkovik, M.: Collective decision making with judgment aggregation. In: Computational Decision Making and Data Science Workshop (CDMDSW 2018), Belgrade, Serbia (2018). http://cdmdsw2018.fon.bg.ac.rs/wp-content/uploads/2019/05/marija_slavkovik_collective_decision_making_with_judgment_aggregation.pdf. Accessed 22 Dec 2019
10. Chen, L., Xu, P., Liu, D.: The effect of crowd voting on participation in crowdsourcing contests. Working paper (2019). 39 pages
11. Bennet, A., Bennet, D.: The decision-making process for complex situations in a complex environment. In: Burstein, F., Holsapple, C.W. (eds.) Handbook on Decision Support Systems 1, pp. 3–20. Springer, Heidelberg (2008). https://doi.org/10.1007/978-3-540-48713-5_1
12. Yu, C., Chai, Y., Liu, Y.: Literature review on collective intelligence: a crowd science perspective. Int. J. Crowd Sci. **2**(1), 64–73 (2018)
13. Mandic, K., Bobar, V., Delibašić, B.: Modeling interactions among criteria in MCDM methods: a review. In: Delibašić, B., et al. (eds.) ICDSST 2015. LNBIP, vol. 216, pp. 98–109. Springer, Cham (2015). https://doi.org/10.1007/978-3-319-18533-0_9
14. Rossi, F.: Preferences, constraints, uncertainty, and multi-agent scenarios. In: ISAIM (2008)
15. Jackson, S.E.: The consequences of diversity in multidisciplinary work teams. In: West, M.A. (ed.) Handbook of Work Group Psychology, pp. 53–75. Wiley, Chichester (1996)
16. Miller, N.R.: Reflections on Arrow's theorem and voting rules. Public Choice **179**, 113–124 (2019)
17. Herings, P.J., Houba, H.: The Condorcet paradox revisited. Soc. Choice Welfare **47**, 141–186 (2016)
18. Dodevska, Z.A.: Computational social choice and challenges of voting in multi-agent systems. Tehnika **74**(5), 724–730 (2019)
19. Chevaleyre, Y., Endriss, U., Lang, J., Maudet, N.: A short introduction to computational social choice. In: van Leeuwen, J., Italiano, Giuseppe F., van der Hoek, W., Meinel, C., Sack, H., Plášil, F. (eds.) SOFSEM 2007. LNCS, vol. 4362, pp. 51–69. Springer, Heidelberg (2007). https://doi.org/10.1007/978-3-540-69507-3_4
20. Endriss, U.: Social choice theory as a foundation for multiagent systems. In: Müller, J.P., Weyrich, M., Bazzan, A.L.C. (eds.) MATES 2014. LNCS (LNAI), vol. 8732, pp. 1–6. Springer, Cham (2014). https://doi.org/10.1007/978-3-319-11584-9_1
21. Conte, R., Gilbert, N., Bonelli, G., et al.: Manifesto of computational social science. Eur. Phy. J. Spec. Topics **214**(1), 325–346 (2012)

22. Brandt, F., Conitzer, V., Endriss, U., Lang, J., Procaccia, A.D.: Introduction to computational social choice. In: Brandt, F., Conitzer, V., Endriss, U., Lang, J., Procaccia, A.D (eds.). Handbook of Computational Social Choice, pp. 1–20. Cambridge University Press, New York (2016)
23. Aitamurto, T., Landemore, H., Galli, J.S.: Unmasking the crowd: participants' motivation factors, expectations, and profile in a crowdsourced law reform. Inf. Commun. Soc. **20**(8), 1239–1260 (2017)
24. Goel, A., Krishnaswamy, A. K., Sakshuwong, S., Aitamurto, T.: Knapsack voting for participatory budgeting. ACM Trans. Econ. Comput. (TEAC) **7**(2), (2019)
25. Ghezzi, A., Gabelloni, D., Martini, A., Natalicchio, A.: Crowdsourcing: a review and suggestions for future research. Int. J. Manag. Rev. **20**(2), 343–363 (2017)
26. Isinkaye, F.O., Folajimi, Y.O., Ojokoh, B.A.: Recommendation systems: principles, methods and evaluation. Egyptian Inf. J. **16**(3), 261–273 (2015)
27. Prpić, J., Shukla, P.P., Kietzmann, J.H., McCarthy, I.P.: How to work a crowd: developing crowd capital through crowdsourcing. Bus. Horiz. **58**(1), 77–85 (2015)
28. Aitamurto, T.: Crowdsourcing for democracy: a new era in policy-making. Publications of the Committee for the Future, Parliament of Finland 1/2012. Helsinki, Finland (2012)
29. Botan, S., Novaro, A., Endriss, U.: Group manipulation in judgment aggregation. In: Thangarajah, J., Tuyls, K., Jonker, C., Marsella, S. (eds.). Proceedings of the 15th International Conference on Autonomous Agents and Multiagent Systems (AAMAS 2016), pp. 411–419, Singapore (2016)
30. Dey, P., Misra, N., Narahari, Y.: Frugal bribery in voting. Theoret. Comput. Sci. **676**, 15–32 (2017)
31. Hata, K., Krishna, R., Fei-Fei, L., Bernstein, M.: A glimpse far into the future: understanding long-term crowd worker quality. In: Proceedings of the 2017 ACM Conference on Computer Supported Cooperative Work and Social Computing, pp. 889–901. ACM DL, Portland (2017)
32. Livshits, B., Mytkowicz, T.: Saving money while polling with InterPoll using power analysis. In: Proceedings of the Second AAAI Conference on Human Computation and Crowdsourcing (HCOMP 2014), pp. 159–170. AAAI Publications (2014)
33. Gaikwad, S.N.S., et al.: Boomerang: rebounding the consequences of reputation feedback on crowdsourcing platforms. In: Proceedings of the 29th Annual Symposium on User Interface Software and Technology, ACM DL, Tokyo, Japan, pp. 625–637 (2016)
34. Göritz, A.S., Borchert, K., Hirth, M.: Using attention testing to select crowdsourced workers and research participants. Soc. Sci. Comput. Rev. (2019)
35. Daniel, F., Kucherbaev, P., Cappiello, C., Benatallah, B., Allahbakhsh, M.: Quality control in crowdsourcing: a survey of quality attributes, assessment techniques, and assurance actions. ACM Comput. Surv. (CSUR) **51**(1) (2018). Article 7, 40 p.
36. Shah, N., Zhou, D.: No oops, you won't do it again: mechanisms for self-correction in crowdsourcing. In: Proceedings of The 33rd International Conference on Machine Learning (ICML), vol. 48, pp. 1–10. New York (2016)
37. Kamangar, Z.U., Kamangar, U.A., Ali, Q., Farah, I., Nizamani, S., Ali, T. H.: To enhance effectiveness of crowdsource software testing by applying personality types. In: Proceedings of the 8th International Conference on Software and Information Engineering, pp. 15–19, Cairo, Egypt. ACM DL (2019)
38. Colman, D.E., Vineyard, J., Letzring, T.D.: Exploring beyond simple demographic variables: differences between traditional laboratory samples and crowdsourced online samples on the Big Five personality traits. Personality Individ. Differ. **133**, 41–46 (2018)
39. Naroditskiy, V., Rahwan, I., Cebrian, M., Jennings, N.R.: Verification in referral-based crowdsourcing. PLoS ONE **7**(10), e45924 (2012)

40. Naroditskiy, V., Stein, S., Tonin, M., Tran-Thanh, L., Vlassopoulos, M., Jennings, N.R.: Referral incentives in crowdfunding. In: Proceedings of the Second AAAI Conference on Human Computation and Crowdsourcing (HCOMP 2014), pp. 171–183. AAAI (2014)

41. Hung, N.Q.V., Thang, D.C., Weidlich, M., Aberer, K.: Minimizing efforts in validating crowd answers. In: Proceedings of the 2015 ACM SIGMOD International Conference on Management of Data, Melbourne, Victoria, Australia, pp. 999–1014. ACM DL (2015)

42. Aitamurto, T., Landemore, H.E.: Five design principles for crowdsourced policymaking: assessing the case of crowdsourced off-road traffic law in Finland. J. Soc. Media Organ. 2(1), 1–19 (2015)

43. Chen, L., Huang, Z., Liu, D.: Pure and hybrid crowds in crowdfunding markets. Financ. Innovation 2, 19 (2016)

44. Aziz, H., Brandt, F., Elkind, E., Skowron, P.: Computational social choice: the first ten years and beyond. In: Steffen, B., Woeginger, G. (eds.) Computing and Software Science. LNCS, vol. 10000, pp. 48–65. Springer, Cham (2019). https://doi.org/10.1007/978-3-319-91908-9_4

45. Faliszewski, P., Skowron, P., Slinko, A., Talmon, N.: Committee scoring rules: axiomatic classification and hierarchy. In: Kambhampati, S. (ed.), Proceedings of the 25th International Joint Conference on Artificial Intelligence, Palo Alto, California USA, pp. 250–256, AAAI Press (2016)

46. Endriss, U.: Computational social choice: prospects and challenges. Procedia Comput. Sci. 7, 68–72 (2011)

47. Gonzalez-Fernandez, Y., Hamidi, S., Chen, S., Liaskos, S.: Efficient elicitation of software configurations using crowd preferences and domain knowledge. Autom. Software Eng. 26, 87–123 (2019)

48. Moreno-Jiménez, J.M., Polasek, W.: E-democracy and knowledge. A multicriteria framework for the new democratic era. J. Multi-criteria Decis. Anal. 12, 163–176 (2003)

49. Haan, M.A., Dijkstra, S.G., Dijkstra, P.T.: Expert judgment versus public opinion–evidence from the Eurovision Song Contest. J. Cult. Econ. 29(1), 59–78 (2005)

50. Mollick, E., Nanda, R.: Wisdom or madness? Comparing crowds with expert evaluation in funding the arts. In: Management Science Articles in Advance, ©2015 INFORMS, pp. 1–21 (2015)

51. Knecht, K., Stefanescu, D.A., Koenig, R.: Citizen Engagement through design space exploration: integrating citizen knowledge and expert design in computational urban planning. In: Sousa, J.P., Castro Henriques, G., Xavier, J.P. (eds.) Architecture in the Age of the 4th Industrial Revolution: eCAADe SIGraDi 2019, vol. 1, pp. 785–794, eCAADe; SIGraDi; FAUD (2019)

52. Drapeau, R., Chilton, L. B., Bragg, J., Weld, D.S.: Microtalk: using argumentation to improve crowdsourcing accuracy. In: Fourth AAAI Conference on Human Computation and Crowdsourcing, September 2016

53. Shams, B., Haratizadeh, S.: Graph-based collaborative ranking. Expert Syst. Appl. 67, 59–70 (2017)

54. Shams, B., Haratizadeh, S.: Reliable graph-based collaborative ranking. Inf. Sci. 432, 116–132 (2018)

Using Multi-level DEA to Go Beyond the Three Dimensions of Sustainability

Georgios Tsaples[⊠] and Jason Papathanasiou

Department of Business Administration, University of Macedonia, Egnatia Str. 156, 54636 Thessaloniki, Greece
gtsaples@gmail.com, jasonp@uom.edu.gr

Abstract. Sustainable development and sustainability are two notions that despite their importance for policy-making, have neither unified definitions nor a common methodological framework to measure them. The purpose of this paper is to propose a measure of sustainability that goes beyond the traditional, three-dimensional structure and accounts also for different definitions and perceptions. To do so, a new, multi-level DEA variation is used, and the method is tested to measure the sustainability of EU countries under three combinations of inputs and outputs (representing perceptions of what sustainability is). The results illustrate that there are countries with variations in the results for the different combinations of inputs and outputs. Thus, no safe conclusions can be drawn for these countries, since the different perceptions alter the measurement level of sustainability.

Keywords: Sustainability · Data Envelopment Analysis · Multi-dimensional structure · Composite indicator

1 Introduction

The term of sustainable development made its appearance in the management and policy-making field in the 1980s with the Brundtland report. It resulted from the recognition that human activities were a constant source of unbalance to natural ecosystems posing severe threats to the security of human societies [1].

Consequently, the report defined sustainable development as the ability and obligation to address the needs of the present without compromising the ability of future generations to do the same [2]. The report and subsequent literature argued that sustainable development could only be achieved if environmental deterioration was deemed equally important as the one of human development and poverty, while suggesting that both issues had to be addressed in a mutual way.

Furthermore, Kates et al. [3, p. 641] indicated that: "the concept of sustainable development does imply limits - not absolute limits, but limitations imposed by the present state of technology and social organization on environmental resources and by the ability of the biosphere to absorb the effects of human activities". Finally, sustainable development should always attempt to exploit technology [4] without however removing it from the cultural context and value system in which it is applied [5].

© Springer Nature Switzerland AG 2020
J. M. Moreno-Jiménez et al. (Eds.): ICDSST 2020, LNBIP 384, pp. 15–29, 2020.
https://doi.org/10.1007/978-3-030-46224-6_2

The above characterizations of sustainable development indicate that its achievement is an enormous, complex effort; nonetheless the first step should always be how to measure it properly [6].

The notion that has been used for that purpose is the one of *sustainability*. It originates from the field of ecology and in its most basic form it signals the ability of a natural system to retain its essential, intrinsic properties and naturally replenish its population. Hence, sustainability is a measure of endurance of natural systems [7].

Despite their importance and extensive use in policy-making, both sustainable development and sustainability are characterized by many definitions, sometimes complementary and sometimes contradictory [4]. In general, all the definitions fall under two categories: in the first, sustainability is seen as a three-dimensional construct, integrating economic, social and environmental dimensions. The second category emphasizes the relationship between humans and nature and underlines the importance of technology as the means that could lead to sustainable development [8].

The lack of a unified definition notwithstanding, the importance of sustainability was immediately understood. However, no methodological guidance or unified framework on how to measure it in practice was offered [9]. One method that has been suggested in the literature and is being continuously used for that purpose is Data Envelopment Analysis (DEA). Still, its use does not come without debate; the main points of contrition are which combinations of inputs and outputs should be used to better capture the notion of sustainability and which variation of DEA is the most appropriate.

Consequently, the objective of the current paper is twofold:

- To propose a measure of sustainability that attempts to integrate both categories of definitions
- To do so by employing a new variation of DEA that mitigates the methodological shortcomings of the method.

The rest of the paper is organized as follows: Sect. 2 is focused on the methodological aspects of the proposed approach. In Sect. 3, results are presented, while in Sect. 4 conclusions and future research directions are discussed.

2 Methodology

The section is divided into three parts. The first offers a glimpse at the literature regarding DEA and sustainability, the second presents the DEA variation, while the third is focused on the data that will be used.

2.1 Sustainability and Data Envelopment Analysis

Zhou et al. [7] performed an extensive literature review on how DEA has been used in relation to sustainability for the years until 2016. Important findings of that effort include the realization that DEA has been increasingly used in sustainability studies especially for the years after 2010. Furthermore, the majority of the papers adopt the three-dimensional structure of sustainability and moreover rely primarily on the economic and environmental dimensions with few studies incorporating the social aspect.

Table 1. Summary of the new research on the literature

Paper	Input	Intermediate	Output	Index	DEA variation	Combination with other method	Area of application
[10]	Labor productivity, capital productivity, the weight of fossil energy and the share of renewable energy in GDP	–	GDP/GHG	Efficiency	Classic DEA	Quantile regression	EU countries
[11]	Consumption of electricity, consumption of heat, consumption of fuel, consumption of sawn wood and particle boards, consumption of fiberboard, consumption of sheets of float glass, consumption of paper and cardboard, consumption of cement, consumption of basic chemicals and plastics, consumption of metallurgical products, water consumption, wastewater discharged in waters, emissions of air pollutants, waste production	–	GDP, gross value added	Eco-efficiency	Classic DEA	–	Polish regions

(continued)

Table 1. (*continued*)

Paper	Input	Intermediate	Output	Index	DEA variation	Combination with other method	Area of application
[12]	Capital, Labor, Energy	–	Gross Regional Product, CO_2 emissions, SO_2 emissions, soot, wastewater, Chemical Oxygen Demand, NO	Efficiency under natural and managerial disposability	Intermediate DEA	–	Chinese regions
[13]	Capital, Labor, Energy	–	Gross Regional Product, CO_2 emissions, SO_2 emissions, soot, waste water, Chemical Oxygen Demand, NO emissions	Efficiency under natural and managerial disposability	Intermediate DEA	–	Chinese region
[14]	AROPE rate, unemployment rate, LCA result, Public school vacancies, number of crimes, inhabitants with higher education	–	Net disposable income	Efficiency	Classic DEA	Material Flow Analysis + Life Cycle Assessment	Spanish cities
[15]	Population, investment in energy industry	Coal consumption, oil consumption, electricity consumption, natural gas consumption	CO_2 emissions, GDP	Efficiency	Two-stage DEA	–	Chinese regions

(*continued*)

Table 1. *(continued)*

Paper	Input	Intermediate	Output	Index	DEA variation	Combination with other method	Area of application
[16]	GDP, population density, labor productivity, total resource productivity, patent applications per 10000 inhabitants	–	GDP per capita, CO_2 emissions	Efficiency	Classic DEA	Malmquist index	German and English cities
[17]	Greenhouse gases, Gross final energy consumption, renewable energy consumption	–	GDP, population	Efficiency	Classic DEA	Zero Sum Gains DEA	EU countries
[18]	Total material consumption, labor unemployment	–	GDP per capita, CO_2 emissions, employment protection index	Efficiency	SORM DEA	Inverse SORM DEA	OECD Countries
[19]	Electricity consumption, total primary energy consumption	–	GDP, GDP per capita, total CO_2 emissions, CO_2/total primary energy	Efficiency and natural and managerial disposability	Intermediate DEA	–	Asian nation
[20]	Mathematical programming scores and scores from the energy trilemma	–	Energy consumption, GHG generations, share of renewable energy in gross final energy consumption	Efficiency	Classic DEA	Mathematical Programming	EU countries
[21]	Employment, Total Energy Consumption, Fixed capital input	–	Total discharge of industrial wastewater, Discharge of industrial waste gas, amount of industrial solid waste	Efficiency	Ray slack-based DEA	–	Chinese regions

(continued)

Table 1. (*continued*)

Paper	Input	Intermediate	Output	Index	DEA variation	Combination with other method	Area of application
[22]	Total renewable energy potential, network length, total installed power of renewable energy, transformer capacity	–	Gross energy generation from renewable energy, number of consumers, total exports, GDP per capita, HDI, Total energy production, Population, area	Super efficiency	Super efficiency DEA	Tobit regression analysis	Turkish regions
[23]	Capital, labor, build-up land, water, energy	–	Solid waste, household refuse, SO_2 emissions, soot, industrial dust, wastewater, GDP	Eco-efficiency	Classic DEA	–	Chinese region
[24]	Infrastructure, efficiency of the legal system, tourists, high school qualifications, unauthorized buildings	–	Environmental index, GDP per capita	Eco-efficiency	Classic DEA	Malmquist index	Italian Regions
[25]	Capital, Labor, Energy	–	Gross regional product (GRP), CO_2 emissions, SO_2 emissions, soot and dust, wastewater, COD, Ammonia nitrogen	Efficiency	Intermediate DEA		Chinese region
[26]	Labor, capital	–	GDP, ecological reserve deficit	Aggregation of efficiency and anti-efficiency	RAM DEA	–	Various nation

(*continued*)

Table 1. (*continued*)

Paper	Input	Intermediate	Output	Index	DEA variation	Combination with other method	Area of application
[27]	Percentage of people with low income, Carbon emissions, Traffic flow, House Price, Anxiety	–	Happiness, Life Satisfaction, Income of tax Payers	Efficiency	Non radial DEA	Temporal analysis	London boroughs
[28]	Capital, Labor, Energy, RFE %	GDP	Wastewater, waste gas, Solid waste, SHC, SBE, SSSE	Unified efficiency	Parallel DEA models	–	Chinese region
[29]	Imports of goods and services in current US$, total annual freshwater withdrawals in percentage of internal resources, public expenditure per capita in current US$, duration of compulsory education	–	Exports of goods and services in current US$, GNI per capita in current US$, total life expectancy at birth in years, total employment, proportion of seats held by women in national parliaments in percentage, CO_2 emissions, total refugees leaving the country	Efficiency	Classic DEA	–	World countries
[30]	Gross Fixed Capital in PPS, Total Labor Force	GDP per capita in PPS	Share of renewable energy in gross final energy consumption, Greenhouse gas emissions (in CO_2 equivalent), Overall life satisfaction, Satisfaction with living environment, Satisfaction with financial situation, Intramural R&D expenditure for all sectors of the economy	Sustainability index	Multi-stage DEA	–	EU countries

A further research was performed in bibliographic databases for the years after 2016. Table 1 above summarizes the new findings.

The first aspect that can be noticed is that in the last few years, researchers have attempted to introduce greater diversification to the combination of inputs and outputs with the aim to include sub-indicators that represent the social dimension. Moreover, as was mentioned before, the DEA literature, for now, addresses sustainability as the typical three-dimensional structure and only in the work of [30] is there an effort to include sub-indicators that focus on technology as a mean to achieve sustainable development. Finally, regarding the variation of DEA, it can be observed that classic DEA variations are preferred compared to others. Two-stage DEA, despite its advantages, has been used only in [15, 28] and [30].

The present paper will build on the works presented in Table 1 and incorporate the social dimension as an equal pillar to the economic and environmental ones and an effort will be made to incorporate more than one perception of what sustainability is.

2.2 Model Formulation

The proposed DEA variation is based on and an extension of the work of [30] and the general framework is presented in Fig. 1 below.

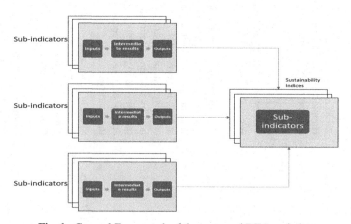

Fig. 1. General Framework of the proposed DEA variation

The main assumption originates from the characteristic of DEA that for any analysis to be meaningful, the number of Decision Making Units under study must be no less than three times of the total number of inputs and outputs [31]. Hence, increasing the number of intermediate stages increases the discriminatory power of the DEA model and the analysis. This assumption applies both for the first stage, where each sub-indicator is calculated separately and in the final indicator of sustainability.

In detail, each dimension of sustainability will be calculated using a two-stage DEA variation. Consequently, the three sub-indicators will be used as an input for a Benefit-of-the-Doubt model, which will result in a geometric, composite indicator of sustainability. The following equations illustrate the proposed model.

Each (overall) sub-indicator of the first stage is calculated from the efficiencies of each step as:

$$E_0 = \xi_1 * \frac{\sum_{d=1}^{D} w_d z_{d0} + u^A}{\sum_{i=1}^{m} u_i x_{i0}} + \xi_2 * \frac{\sum_{r=1}^{s} u_r y_{r0} + u^B}{\sum_{d=1}^{D} w_d z_{d0}} \qquad (1)$$

ξ_1, ξ_2 *indicate the weight for each stage*
x_{ij} *indicates the input i of DMU j of stage 1*
z_{dj} *indicates the output d of DMU j of stage 1 or the input of stage 2*
y_{rj} *indicates the output r of DMU j of stage 2*
u_i, w_d, u_r *are the weights for each input and output of each stage*

The additive two-stage model is calculated under Variable Returns to Scale as follows:

$$\max \sum_{d=1}^{D} \mu_d z_{d0} + \sum_{r=1}^{s} \gamma_r y_{r0} + u^1 + u^2 \qquad (2)$$

$$s.t. \sum_{i=1}^{m} \omega_i x_{i0} + \sum_{d=1}^{D} \mu_d z_{d0} = 1 \qquad (3)$$

$$\sum_{d=1}^{D} \mu_d z_{dj} - \sum_{i=1}^{m} \omega_i x_{ij} + u^1 \leq 0, j = 1, 2, \ldots n \qquad (4)$$

$$\sum_{r=1}^{s} \gamma_r y_{rj} - \sum_{d=1}^{D} \mu_d z_{dj} + u^2 \leq 0, j = 1, 2, \ldots n \qquad (5)$$

$$\gamma_r, \mu_d, \omega_i \geq 0, \text{ the optimal multipliers} \qquad (6)$$

$$i = 1 \ldots m, \ j = 1 \ldots n, \ r = 1 \ldots s, \ d = 1 \ldots D, \ u^1, u^2 \text{ free} \qquad (7)$$

After calculating each sub-indicator in the first stage, a Benefit-of-the-Doubt model is used to calculate the optimal weights for the final composite, sustainability indicator. The mathematical formulation of the model is:

$$\max \sum_{r=1}^{s} w_{ri} y_{ri} \qquad (8)$$

$$s.t. \sum_{r=1}^{s} w_{ri} y_{rj} \leq 1 (N \text{ constraints, one for each DMU } j = 1 \ldots N) \qquad (9)$$

$$w_{rj} \geq 0, (s \text{ constraints one for each sub} - indicator) \qquad (10)$$

Finally, after calculating the optimal weights for each sub-indicator. the composite indicator of sustainability is calculated by:

$$CI_j = \prod_{r=1}^{s} \left(\frac{y_{rj}}{y_{rb}} \right)^{\omega_r^*} \qquad (11)$$

Where y_{rB} are considered base values on which the numerators are compared, thus keeping the composite indicator unit invariant. Furthermore, the values of ω_r^* indicate how much the rth sub-indicator contributes to the final sustainability index. The process is repeated as many times as the different perceptions of sustainability that are wished to be calculated.

2.3 Data and Sub-indicators

As it was mentioned before, the majority of the studies relies on the three-dimensional structure of sustainability. However, even within this approach, different combinations of inputs and outputs are used for the DEA models. This is to be expected, since aspects such as the social dimension or what is the most important elements of the environmental dimensions, could have different meanings for different people.

Nonetheless, by using one combination of inputs and outputs, other perceptions (where perception means a specific set of inputs and outputs) are excluded from the analysis. To address the specific issue, in the current paper, three indices of sustainability will be calculated, representing different perceptions. The sub-indicators that will be used are:

- Economic-Environmental

  - Inputs: Gross Fixed Capital in Purchasing Power Standards (PPS), Total Labor Force
  - Outputs/Second stage input: GDP per capita in PPS
  - Outputs: Share of renewable energy in gross final energy consumption, Greenhouse gas emissions (in CO_2 equivalent)

- Economic-Social

  - Inputs: Gross Fixed Capital in PPS, Total Labor Force
  - Outputs/Second stage input: GDP per capita in PPS
  - Outputs: Overall life satisfaction, Satisfaction with living environment, Satisfaction with financial situation

- Economic-Research and Development

  - Inputs: Inputs: Gross Fixed Capital in PPS, Total Labor Force
  - Outputs/Second stage input: GDP per capita in PPS
  - Outputs: Intramural R&D expenditure for all sectors of the economy

- Economic

  - Inputs: Gross Fixed Capital in PPS, Total Labor Force
  - Outputs/Second stage input: GDP per capita in PPS
  - Outputs: Mean equivalized net income, ability to face unexpected financial expenses as percentage of the population

Table 2 summarizes the three indices of sustainability that will be calculated. In the second column, the sub-indicators that will be combined are displayed for each sustainability index and finally each sub-indicator will result from the combination of inputs and outputs mentioned above.

Table 2. The three indices of sustainability and their sub-indicators

Sustainability index	Sub-indicators
Sustainability index 1	Economic-environmental, economic-social, economic-research and development
Sustainability index 2	Economic, economic-environmental, economic-social
Sustainability index 3	Economic, economic-environmental, economic-social, economic-research and development

The data were taken by Eurostat [32] and the area of application will be for the EU countries. The inputs and outputs for each sub-indicator were sampled by the works presented in Table 1 and they are being used merely as an illustration for the proposed method.

3 Results

The first part of the analysis is focused on the calculation of the sub-indicators based on the raw data. The following Table 3 illustrates the results from the first part of the analysis.

Table 3. Results for the sub-indicators for the EU countries

Country	Econ-Environmental	Econ-Social	Economic-R&D	Economic
Belgium	0.211	0.381	0.387	0.474
Bulgaria	0.545	0.545	0.545	0.545
Czech Republic	0.291	0.367	0.370	0.4
Denmark	0.250	0.542	0.501	0.54
Germany	0.213	0.335	0.419	0.45
Estonia	0.487	0.487	0.522	0.58
Ireland	0.226	0.419	0.343	0.47
Greece	0.352	0.386	0.388	0.44
Spain	0.262	0.303	0.335	0.45
France	0.215	0.327	0.386	0.48
Croatia	0.512	0.105	0.467	0.52
Italy	0.238	0.271	0.324	0.443
Cyprus	0.588	0.617	0.608	0.8
Latvia	0.608	0.538	0.493	0.56

(continued)

Table 3. (*continued*)

Country	Econ-Environmental	Econ-Social	Economic-R&D	Economic
Lithuania	0.579	0.579	0.421	0.46
Luxemburg	0.588	0.674	0.7	1
Hungary	0.365	0.401	0.407	0.4
Malta	0.770	0.817	0.803	1
Netherlands	0.186	0.443	0.331	0.42
Austria	0.244	0.523	0.423	0.48
Poland	0.349	0.506	0.371	0.41
Portugal	0.363	0.356	0.387	0.43
Romania	0.511	0.511	0.436	0.43
Slovenia	0.394	0.522	0.512	0.55
Slovakia	0.335	0.384	0.361	0.41
Finland	0.309	0.537	0.524	0.52
Sweden	0.520	0.472	0.520	0.52
United Kingdom	0.216	0.384	0.340	0.46

The above sub-indicators are combined in the ways that were described on Table 2 to produce the three (geometric) composite indices of sustainability. Figure 2 below presents the results.

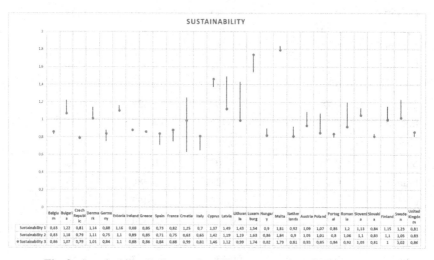

Fig. 2. Sustainability indices under different perceptions for EU countries

The first result to be observed is that Malta and Luxemburg have the highest values of sustainability. Second, there are countries that show very little variation in the value of the sustainability index under the different perceptions. For example, Malta, Belgium and the Czech Republic. On the other hand, there are countries with observable variation in the results for the three sustainability indices. Examples include Luxemburg, Poland, and Romania. Finally, Croatia is the country with the largest variation in the results. Consequently, no safe conclusions can be drawn for the specific country since depending on the perception (meaning the set of inputs and outputs used) Croatia could have the lowest level of sustainability or one of the highest.

In conclusion, the results illustrate that altering the perception of what sustainability is could alter the results of its measurement. This fact may be self-evident, but it could serve as a crucial point for policy makers since it can highlight why the different perceptions result in different measurements, thus indicating potential policies that could increase the overall sustainability, without acting in the expense of areas that are in a good direction.

4 Conclusions

The purpose of the current paper was to propose a new measure of sustainability by integrating the various definitions of sustainability. To do so, a multi-level Data Envelopment Analysis method was used.

A review in the literature indicated that studies that employ DEA to measure sustainability use the three-dimensional approach in their majority. However, in that way other perceptions are eliminated. Furthermore, classic variations of DEA are used despite the fact that a network approach could increase the discriminatory power of the analysis and mitigate some of the limitations of the classic DEA methods.

The proposed method was used to measure the sustainability of EU countries under three different perceptions (combinations of inputs and outputs) that represent different definitions of sustainability. The results indicated that there are countries that have consistent levels of sustainability (regardless of the combination of inputs and outputs) and others that present variations, making difficult the extraction of robust conclusions.

Future research directions of the research include the incorporation of even more and diverse perceptions of sustainability along with different equations to calculate the index of sustainability. Furthermore, the research was limited in the field of DEA and did not consider attempts from different areas of research such as the work by Hay et al. [33]. Finally, all these results (and different approaches) could be analyzed with the help of Artificial Intelligence algorithms that could also automate the generation of the results.

References

1. Coli, M., Nissi, E., Rapposelli, A.: Monitoring environmental efficiency: an application to Italian provinces. Environ. Model. Softw. **26**(1), 38–43 (2011)
2. Brundtland, G.: Report of the World Commission on environment and development: our common future, United Nations (1987)
3. Kates, R., et al.: Sustainability science. Science **292**(5517), 641–642 (2001)
4. Robinson, J.: Squaring the circle? Some thoughts on the idea of sustainable development. Ecol. Econ. **48**(4), 369–384 (2004)

5. Santana, N., Mariano, E., Camioto, F., Rebelatto, D.: National innovative capacity as determinant in sustainable development: a comparison between the BRICS and G7 countries. Int. J. Innov. Sustain. Dev. **9**(3–4), 384–405 (2015)
6. Tyteca, D.: Sustainability indicators at the firm level. J. Ind. Ecol. **2**(4), 183–197 (1998)
7. Zhou, H., Yang, Y., Chen, Y., Zhu, J.: Data envelopment analysis application in sustainability: the origins, development and future directions. Eur. J. Oper. Res. **264**(1), 1–16 (2018)
8. Drucker, P.: Innovation and Entrepreneurship. Routledge, Abingdon (2014)
9. Munda, G., Saisana, M.: Methodological considerations on regional sustainability assessment based on multicriteria and sensitivity analysis. Reg. Stud. **45**(2), 261–276 (2011)
10. Moutinho, V., Madaleno, M., Robaina, M.: The economic and environmental efficiency assessment in EU cross-country: evidence from DEA and quantile regression approach. Ecol. Ind. **78**, 85–97 (2017)
11. Masternak-Janus, A., Rybaczewska-Błażejowska, M.: Comprehensive regional eco-efficiency analysis based on data envelopment analysis: the case of Polish regions. J. Ind. Ecol. **21**(1), 180–190 (2017)
12. Sueyoshi, T., Yuan, Y.: Social sustainability measured by intermediate approach for DEA environmental assessment: Chinese regional planning for economic development and pollution prevention. Energy Econ. **66**, 154–166 (2017)
13. Sueyoshi, T., Toshiyuki, Y., Li, A., Wang, D.: Methodological comparison among radial, non-radial and intermediate approaches for DEA environmental assessment. Energy Econ. **67**, 439–453 (2017)
14. Gonzalez-Garcia, S., Manteiga, R., Moreira, M., Feijoo, G.: Assessing the sustainability of Spanish cities considering environmental and socio-economic indicators. J. Clean. Prod. **178**, 599–610 (2018)
15. Lin, T., Chiu, S.: Sustainable performance of low-carbon energy infrastructure investment on regional development: evidence from China. Sustainability **10**(12), 4657 (2018)
16. Moutinho, V., Madaleno, M., Robaina, M., Villar, J.: Advanced scoring method of eco-efficiency in European cities. Environ. Sci. Pollut. Res. **25**(2), 1637–1654 (2018). https://doi.org/10.1007/s11356-017-0540-y
17. Cucchiella, F., D'Adamo, I., Gastaldi, M., Miliacca, M.: Efficiency and allocation of emission allowances and energy consumption over more sustainable European economies. J. Clean. Prod. **182**, 805–817 (2018)
18. Hassanzadeh, A., Yousefi, S., Farzipoor Saen, R., Hosseininia, S.S.S.: How to assess sustainability of countries via inverse data envelopment analysis? Clean Tech. Environ. Policy **20**(1), 29–40 (2017). https://doi.org/10.1007/s10098-017-1450-x
19. Sueyoshi, T., Yuan, Y.: Measuring energy usage and sustainability development in Asian nations by DEA intermediate approach. J. Econ. Struct. **7**(6), 1–18 (2018). https://doi.org/10.1186/s40008-017-0100-0
20. Biresselioglu, M., Demir, M., Turan, U.: Trinity on thin ice: Integrating three perspectives on the European Union's likelihood of achieving energy and climate targets. Energy Res. Soc. Sci. **42**, 247–257 (2018)
21. Song, M., Peng, J., Wang, J., Zhao, J.: Environmental efficiency and economic growth of China: a ray slack-based model analysis. Eur. J. Oper. Res. **269**(1), 51–63 (2018)
22. Ervural, B., Zaim, S., Delen, D.: A two-stage analytical approach to assess sustainable energy efficiency. Energy **164**, 822–836 (2018)
23. Yang, L., Zhang, X.: A comprehensive eco-efficiency model and dynamics of regional eco-efficiency in China. J. Clean. Prod. **173**, 100–111 (2018)
24. Carboni, O., Russu, P.: Measuring and forecasting regional environmental and economic efficiency in Italy. Appl. Econ. **50**(4), 335–353 (2018)

25. Zhang, A., Li, A., Gao, Y.: Social sustainability assessment across provinces in China: an analysis of combining intermediate approach with data envelopment analysis (DEA) window analysis. Sustainability **10**(3), 732 (2018)
26. DiMaria, C.-H.: An indicator for the economic performance and ecological sustainability of nations. Environ. Model. Assess. **24**(3), 279–294 (2018). https://doi.org/10.1007/s10666-018-9626-2
27. Pozo, C., Limleamthong, P., Guo, Y., Green, T., Shah, N., Acha, S., Guillén-Gosálbez, G.: Temporal sustainability efficiency analysis of urban areas via Data Envelopment Analysis and the hypervolume indicator: application to London boroughs. J. Clean. Prod. **239**, 1–14 (2019)
28. Zhao, L., Zha, Y., Zhuang, Y., Liang, L.: Data envelopment analysis for sustainability evaluation in China: tackling the economic, environmental and social dimensions. Eur. J. Oper. Res. **275**(3), 1083–1095 (2019)
29. Tajbakhsh, A., Shamsi, A.: Sustainability performance of countries matters: a non-parametric index. J. Clean. Prod. **224**, 506–522 (2019)
30. Tsaples, G., Papathanasiou, J., Georgiou, Andreas C., Samaras, N.: Assessing multidimensional sustainability of European countries with a novel, two-stage DEA. In: Freitas, P.S.A., Dargam, F., Moreno, J.M. (eds.) EmC-ICDSST 2019. LNBIP, vol. 348, pp. 111–122. Springer, Cham (2019). https://doi.org/10.1007/978-3-030-18819-1_9
31. Wu, D.: Performance evaluation: an integrated method using data envelopment analysis and fuzzy preference relations. Eur. J. Oper. Res. **194**(1), 227–235 (2009)
32. EUROSTAT, *Database,* Eurostat, 2019
33. Hay, L., Duffy, A.H., Whitfield, R.I.: The S-Cycle performance matrix: supporting comprehensive sustainability performance evaluation of technical systems. Syst. Eng. **20**(1), 45–70 (2017)

Analysis of Graphical Visualizations for Multi-criteria Decision Making in FITradeoff Method Using a Decision Neuroscience Experiment

Lucia Reis Peixoto Roselli$^{(\boxtimes)}$ (iD) and Adiel Teixeira de Almeida (iD)

Center for Decision Systems and Information Development (CDSID),
Universidade Federal de Pernambuco, Recife, PE, Brazil
{lrpr,almeida}@cdsid.org.br

Abstract. The use of bar graphics and tables to represent a Multi-Criteria Decision Making/Aiding (MCDM/A) problems are investigated in this study since these visualizations present a holist vision for MCDM/A situations. In this context, these visualizations bring flexibility to the decision-making process conducted in the Decision Support System (DSS) developed for the FITradeoff method, being an important advantage in this method. In order to support this study, the Neuroscience approach is aggregated to MCDM/A and a neuroscience experiment is constructed to investigate how decision-makers (DMs) evaluate bar graphs and tables in order to identify some patterns of behavior. The main task required in this experiment was to evaluate MCDM/A situations and select the alternative which performed best. Based on descriptive and statistical analyses of the results, some suggestions could be made about DMs behavior´s when the visualizations were evaluated. Therefore, for this study, two main purposes were raised: provide insights for the analyst about the use of graphical and tabular visualization in MCDM/A situations and to improve the FITradeoff Decision Support System. Regarding to the first purpose, a advising rule has been built to support the analyst in the advising process performed with the DMs. Regarding to the second purpose was suggested that tables should be included in the FITradeoff DSS. In total, 51 Management Engineering students took part in the experiment.

Keywords: Multi-Criteria Decision Making/Aiding (MCDM/A) · Graphical visualization · FITradeoff method · Decision Support System

1 Introduction

Multi-Criteria Decision Making/Aiding (MCDM/A) deals with different types of problems which are characterized by a group of alternatives based on some criteria of interest [1–4]. An important task presented in the MCDM/A approach is preference modeling, which is used to identify and elicit Decision Makers (DMs) preferences for the problem.

However preference modeling is not a simple task with tradeoffs often involved. This poses the question if it is better to have simple mathematical models which present

J. M. Moreno-Jiménez et al. (Eds.): ICDSST 2020, LNBIP 384, pp. 30–42, 2020.
https://doi.org/10.1007/978-3-030-46224-6_3

distortions in the calculations but are relatively easy for the DMs to understand and which take account of their preferences or to have robust mathematical models which have a robust structure but which are difficult for the DMs to understand and which take note of their preferences. For these two cases, the solutions generated cannot represent the DM's real preferences, as result from distortions during the preference modeling process.

An example is presented in the Traditional Tradeoff elicitation procedure [1] which presents a robust axiomatic structure. However, the DM need to indicate indifference relations for each pair of criteria in order to find the exact value of scaling constants which is considered complex and can lead to 67% of inconsistencies in the solutions [5].

On the other hand, the FITradeoff method [6], uses the same axiomatic structure of the Tradeoff elicitation procedure, although with partial information to elicit scaling constant. Compared to the Tradeoff, in FITradeoff, the DMs present only their strict preferences, thus the weight space is found based on intervals of values for each scaling constant.

Therefore, in order to investigate an important feature presented in the FITradeoff Decision Support System (DSS), the holist evaluations process, which is performed using graphical visualization in FITradeoff DSS, this study has been developed.

The Graphical Visualization has been the focus of this study because this tool provides a holist evaluation of Potential Optimal Alternatives (POA) in the MCDM/A problem, bringing flexibility to the decision-making process. The aim of this study is to investigate how the DM evaluates MCDM/A situations represented by bar graphics and tables in order to identify some patterns of behavior to improve the holistic evaluation phase presented in FITradeoff DSS and to provide insights to the analyst during the advising process performed with DMs.

In this context, in order to support the behavior investigation the Neuroscience approach was used, being an important supplement to investigate the decision-making process [7].

The use of Neuroscience approach to investigate the decision-making process is discussed in the literature in many contexts, such as: in economics – Neuroeconomics [8, 9], in marketing – Consumer Neuroscience and Neuromarketing [10, 11], and in information systems – NeuroIS [12, 13].

According to [14], regarding to the combination of Neuroscience with MCDM/A studies on decision-making in the context of MCDM/A is not very integrated with the Neuroscience approach. For this theme, there are a small number of studies in the literature at the moment [14–17].

Therefore, in order to explore this theme a neuroscience experiment was applied using an Eye-Tracker and Electroencephalogram (EEG) to capture DM physiologic variables and investigate how DM evaluates bar graphs and tables which represent MCDM/A situations.

This paper is organized as follows. Section 2 describes the FITradeoff method, Sect. 3 presents the Neuroscience Sect. 4 presents the Results found in the analyses and Sect. 5 the discussion of these results. Finally Sect. 6 draws some Conclusions and suggests lines for future studies.

2 FITradeoff Method

The Flexible and Interactive Tradeoff (FITradeoff) Method [6] was developed to elicit scaling constants for MCDM/A context. This method is in the context of Multi-Attribute Value Theory – MAVT [1, 2] and presents the same axiomatic structure as the Tradeoff Method [1].

In order to explain the FITradeoff method, a hypothetical supplier selection problem with five alternatives evaluated in four criteria is solved, which the consequences presented in Table 1. A Decision Support System was developed to conduct the FITradeoff process and it is available by request at www.fitradeoff.org.

Table 1. Supplier selection decision matrix

Alternative vs criteria	Price	Capacity	Reliability	Location
Criterion description	In dollars ($)	In units	Where 3 is good, 2 is regular and 1 is bad	Where 3 is good, 2 is regular and 1 is bad
Supplier 1	60	40	3	3
Supplier 2	50	80	2	2
Supplier 3	40	60	1	3
Supplier 4	30	55	1	1
Supplier 5	50	65	2	2

Regarding how to apply the FITradeoff method to solve this selection problem, in the first step of the FITradeoff method the DMs have to rank the criteria scaling constant or weights considering their preferences about the consequences presented for each criterion.

Therefore, for this situation, considering the DM preferences, the ranking of criteria weights is defined as Price Weight > Capacity Weight > Reliability Weight > Location Weight. Based on this order, the first inequality is created, being inserted in the Linear Programing Problem (LPP) in order to seek for a solution.

In this context, after the updating of the LPP model, some alternatives were dominated from other, being the number of alternatives reduced from five to three, i.e. the alternatives Supplier 1 and Supplier 5 were eliminated of this hypothetical supplier selection decision-making process exemplified in this study to demonstrate how the FITradeoff method works.

In Step 2 the elicitation process is performed in order to reduce the weights space. In this step some questions are presented to the DMs, who have to express only their strict preferences about the evaluation of some alternatives in some criterion, i.e. the consequences. Figure 1 illustrates one elicitation question which asks the DM to compare the consequence of having 50($) in price or the best performance (80 units) in Capacity. The other criteria are at worse values because in this method the criteria are compared in pairs. For this elicitation question the DM prefers Consequence A, i.e. having a supplier with 50 in price. In this method the DMs are present throughout the whole decision-making process presenting their preferences.

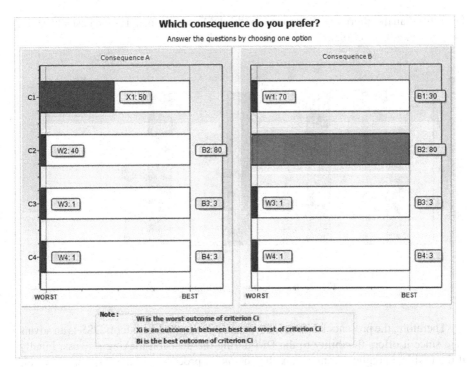

Fig. 1. Elicitation question

The FITradeoff method provides, in parallel with the elicitation process (step 2), the graphical visualization phase. Thus, some graphics are presented in the DSS to support the DMs in the understanding of the problem. Thus, if they wish a final alternative can be selected in the group of Potentially Optimal Alternatives (POA), before the final alternative is defined by the LPP model. The FITradeoff DSS presents three types of graphics in the graphical visualization phase: a bar graph, a bubble graph and a spider graph to support the DM to solve MCDM/A situations.

For this illustrative supplier selection example, after the DM answered 4 elicitation questions, the number of alternatives was reduced from three to two (the alternative Supplier 3 was eliminated). Therefore. for this supplier selection situation exemplified in this paper, when the number of alternatives was reduced to two, the DM considered comfortable to evaluate these two POAs using the graphical visualization form.

Hence, based on the bar graph presented in Fig. 2, the DM observes that alternative Supplier 4 (highlighted in yellow) presented the best value in price (with the lowest price in the problem). However, regarding the other criteria, the alternative Supplier 4 presented unfavorable evaluations (consequences) than the alternative Supplier 2 (highlighted in blue), with a difference below 50% in performance in the criterion Reliability and the criterion Localization. Also, based in the evaluation of the difference in performance for the criterion Price, in which the Supplier 2 presented 60% of performance in Price and the best performance in the other three criteria, the alternative Supplier 2 is considered favorable than the alternative Supplier 4 for this DM and is selected as the best alternative for this illustrative problem.

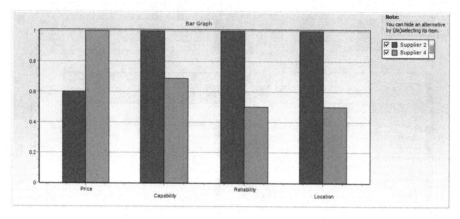

Fig. 2. Bar graphic (Color figure online)

Therefore, the presence of graphical visualization in the FITradeoff DSS is an advantage since it offers flexibility to the DM during the decision-making process. Finally, if the DM had continued answering the elicitation process, another 2 questions would have been presented before the alternative Supplier 2 would have been indicated as the best alternative for this problem. It is worth pointing out that this number of questions is not fixed; it depends on the preferences expressed by the DM.

3 Neuroscience Experiment to Investigate Graphical Visualization in DSS

The neuroscience experiment was performed to use neuroscience tools as a supplement to investigate DM behavior when graphical visualization was used to represent MCDM/A problems.

For this study, bar graphics and tables were compared based on the suggestion presented in the previous study which concluded that tables performed positively to represent MCDM/A situations in the FITradeoff context [18–20].

To construct the experiment ten MCDM/A situations were built using the random function in Excel. These problems were constructed presenting consequences between 0 and 1, being normalized by the maximum value presented for each criterion. Also, these problems were constructed using values between 0 and 1 in order to present the same normalization procedure as used in the Tradeoff method [1]. Moreover, these problems were constructed out of a particular context to allow the generalization of the behavioral results. Thus they presented alternatives named A, B and C and criteria named 1, 2, 3.

For these ten MCDM/A situations, which were constructed using three, four and five alternatives evaluated in three, four and five criteria, 10 bar graphs and 10 tables were generated. Thus, pairs of bar graphs and tables were constructed using the same decision matrix, to allow comparisons.

Moreover, five of these problems used the same values for the criteria weights and five of them used different values for criteria weights. For problems that had different weights, the first criterion presented the highest value of the weight and the last criterion presented the lowest value. The weights were calculated following an arithmetic progression and a tendency line was designed in these bar graphs to highlight the difference in weights. Figure 3 presents a graphic named 4A3C BD, i.e. a Bar graphic (B) with Different weights (D), four Alternatives (4A) and three Criteria (3C). Also, considering the table generated from this MCDM/A configuration (four Alternatives and three Criteria with Different weights) the acronym used is 4A3C TD. In situations with same values to the criteria weights the acronyms is BS or TS.

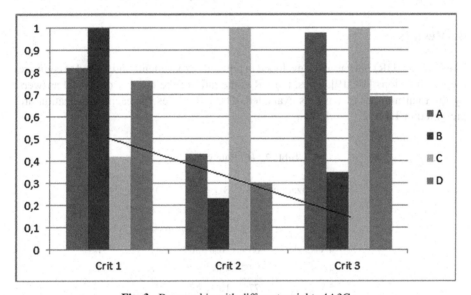

Fig. 3. Bar graphic with different weights 4A3C

To design this experiment the 20 visualizations constructed were displayed in a random sequence, based on a previous study which concluded that the random sequence was the best sequence for displaying the visualizations [18, 19].

Thus, the only task required during the experiment is to evaluate each visualization and select the best alternative, following MAVT concepts [1, 2]. Therefore, after each participant evaluates a table or a bar graph, the question was asked: Which is the best alternative in the visualization?

The experiments took place in the same room in the laboratory of NeuroScience for Information and Decision (NSID) in the Federal University of Pernambuco (Universidade Federal de Pernambuco - UFPE), in Recife – Brazil. Two volunteers participated in each experiment. At the start of the experiment, the volunteers explained the experiment and presented two documents, the first of which gave instructions about the task required while the second was a term of consent. This experiment was approved by the University Ethics Committee.

In total, 51 Management Engineering students from UFPE participated in this experiment. In this sample, there were 25 women and 26 men, of whom 28 were undergraduates and 23 were post-graduate students. For this sample, 37 participants were below 26 years old and 14 were above 26 years old. This experiment was part of students' activities in their course on Decision-Making.

In this experiment two neuroscience tools were used: an X120 Eye-Tracker by Tobbi Studio and the 14 channels EEG by Emotiv. The eye-tracker was used to design the experiment using its software tools. In addition, the answers provided by each participant for each questionnaire were collected by this software and used to perform some analyses. The EEG was used to collect the Alpha and Theta activities in some channels in order to investigate the participant's behavior during the experiment.

4   Results

The Hit Rate (HR) is a variable developed in previous experiments and is used to compare the visualizations [18, 19]. Thus, the HR is the ratio of the number of correct answers by the total number of answers evaluated. The HR values for each visualization are presented in Table 2.

Table 2. Hit Rate values

Visualizations	BS	BD	TS	TD
3A4C	97%	47%	90%	43%
4A3C	33%	53%	33%	63%
4A4C	63%	20%	50%	30%
4A5C	83%	43%	83%	50%
5A4C	77%	10%	67%	17%

In order to investigate the HR values some comparisons can be made using a descriptive analysis. The first group is used to compare same weights and different weights (BS versus BD and TS versus TD). The second group was to compare the bar graphic and table (BS versus TS and BD versus TD).

Based on the first comparison, same weights versus different weights, the HR values suggested that visualizations with the same weights presented a higher HR than visualizations with different weights, either for bar graphics or tables. For the second group, comparison of bar graphics versus tables, the results suggested that bar graphics and tables with same weights presented similar HR values, two of which (4A3C and 4A5C) had the same HR. For bar graphics and tables with different weights, the results indicate that tables were better than bar graphics in four out of five comparisons. In order to continue the investigation of HR values, Theta and Alpha analyses were also performed.

According to [21–24] higher value of Theta Band (4–8 Hz) in frontal channels and the suppression of values in Alpha Band (8–12 Hz) in posterior cortex region can suggest mental activity and engagement. Therefore, in order to identify the difficulty in evaluating some visualization these values are evaluated.

To perform the Theta and Alpha analyses the Wilcoxon Signed Rank Test was performed with $\alpha = 5\%$ [25]. In this context, the visualizations pairs analyzed when comparing HRs, were evaluated again using the statistical analysis of Theta Band in channels F3, F4, F7, F8 and Alpha Band in channels P7, P8, O1, O2. These channels were selected because they are in the frontal and posterior brain area.

In this context, based on the Wilcoxon Signed Rank Test results, it was observed that for Theta activity analysis the hypothesis H0 was rejected for some cases (4A3C, 4A4C and 4A5C), which suggests that for these visualizations the Theta Band values are different, i.e. the hypothesis H0 is that values of in the visualization pair are similar. Therefore, based on the subtraction of these values for these cases the results indicate that the power of Theta Band was higher for BD than BS and for TD than TS.

Moreover, an interesting suggestion of this result is that, for the comparison of bar graphic and tables, the hypothesis H0 was not rejected in all cases, suggesting that, for bar graphics and tables, the complexity in evaluation was similar for the pairs. This result is also coherent with the result for the HR values.

For Alpha activity analysis the hypothesis H0 was rejected for some cases, indicating that similar values in alpha power for some visualization pairs were not observed. Again based on the subtraction of these values, it could be seen that BS presented a higher value in the Alpha Band than TS did. It is worth to mentioning that to perform these analyses the power values are not referenced by an initial activity.

5   Discussion of Results

Based on the analyses developed, the descriptive analysis with HR values and statistical analyses with Theta and Alpha values, the results found were coherent.

For HR was observed that it was more difficult for the participants to evaluate visualizations with different weights than visualizations with the same weights. This result is consistent with the result found based on Theta values, which indicates that the power of Theta Band was higher for BD than BS and for TD than TS, reinforcing the suggestion that different weights is a complex case since these present lower HR values and higher values of potency in the Theta Band than same weights do.

Moreover, for the comparison of tables and bar graphics, based on HR values and Theta analysis, the complexity in evaluation was similar for the pairs, especially for same weights. However, based on Alpha analysis was observed a higher engagement in TS when this form of visualization was compared to BS. This result is important because reinforce the conclusion, presented in the previous study [18, 19], that tables should be included in FITradeoff DSS. For BD and TD, more evaluation need to be done since based on Alpha analysis these visualizations presented similar engagement, but based on HR values tables were better than bar graphics in four out of five comparisons.

Also, the HR values is an important insights generated from the neuroscience experiments because these values are a direct recommendation from the analyst to the DM about the use or not use of the visualization to define an preference relation, i.e. if an alternative is preferable than other [18, 19]. The HR values are assumed to be distributed according to a Bernoulli distribution [25], being HR the percentage of success when only one test was performed, i.e. for each visualization the participants had only one chance to select the best alternative. Therefore, based on this distribution, the variance and the standard deviation could be calculated.

For a general case, presenting a range of 0% to 100% of HR values (or the probability of success 0 to 1), the standard deviation has been computed in order to construct an advising rule to be followed by the analyst when interacting with the DM. The three regions advising rule is constructed in this study as illustrated in Fig. 4, in which the horizontal axis shows the probability of success and the vertical axis shows the standard deviation.

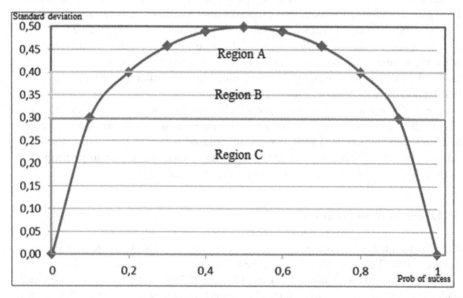

Fig. 4. Three regions advising rule

It can be seen that for extremes probabilities of success, the standard deviation have lesser values. On the other hand, for central probabilities of success, around 0,5 the standard deviation is greater. Therefore, these metrics can support the analysts in the recommendations, since for higher or lower probabilities of success, the analyst can strength advice the DM respectively to accept or reject some visualization based on the small variability of these values. In the other hand, for probabilities of success between 0,4 and 0,6 the analyst can advise the DM to be more cautions and possibility do not use the information given by the visualization, because the variability of the result is comparatively higher.

Moreover, related to Theta and Alpha analyses, an additional analysis can be developed merging Theta and Alpha values. As a result, a quadrant diagram can be constructed from Alpha and Theta sign combination. This diagram is constructed based on the meaning of Alpha and Theta activities in frontal and posterior cortex regions [21–24]. Based on this diagram, five patterns of behavior can be observed, the four patterns illustrated in Fig. 5, and a disperse pattern.

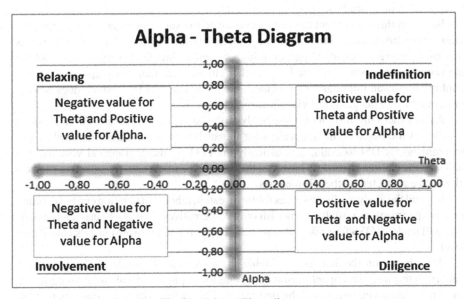

Fig. 5. Alpha vs Theta diagram

These patterns have been observed in the experiment and show that different participants had these five different behaviors. Based on initial investigations was possible to observe that most participants were in a particular quadrant for the majority of the visualizations presented in the experiments. For instance, some participants follow the 'diligent' pattern, meaning that they are engaged with cognitive effort most of time, while others may follow a relax pattern, with neither engaged nor cognitive effort, most of time. On the other hand some participants do not follow one of the four patterns of the quadrant, meaning that they are disperse in their behavior, following one of the four patterns of the quadrant at a time.

6 Conclusion

This study is developed to investigate how the decision-maker evaluated MCDM/A situations represented using bar graphics and tables. The holistic evaluation phase, performed in the FITradeoff method using graphics, is the focus of this study because it is an important feature of the FITradeoff method, which makes the decision-making process more flexible.

Even, being constructed considering the FITradeoff method, the results generated in this study, about the use of bar graphics and tables to represent MCDM/A situations, can be used to support the analyst in the advising process constructed in a decision-making process with other MCDM/A methods, in MAVT context [1] with additive model. This study is not exclusive for the FITradeoff method, being the bar graphics and tables designed with the MCDM/A consequence matrix information (i.e. using the performance of the alternatives evaluated in criteria) which allow that these results can be applied to other studies.

Based on this experiment the results confirmed the suggestion generated in the previous experiment, that tables should be included in the FITradeoff DSS, since tables that used different weights presented a higher number hits than bar graphics, and for the same weight, using tables led to a higher engagement than bar graphics did. This confirmation is an important improvement for the FITradeoff DSS which does not yet present tables in the graphical visualization section.

As to generating insights for the analyst, this study suggested the Three Regions Advising Rule based on the HR values for the analyst. In other words, the analyst can recommend for DM to evaluate holistically the POA by using graphical visualization and skip the elicitation process based on the confidence of success for selecting the best alternative indicated by the HR analysis.

Beyond the analyses discussed, as future research, other visualization can be included in the experiment, as visualizations with only 2 alternatives. This configuration is essential in FITradeoff DSS for rank problems [26], being important to evaluate its confidence level.

Also, based on the analysis of Alpha-Theta Diagram is possible to correlate the pattern found with participant personal information's, physiological variables collected by neuroscience equipment and the performance of results, which are analyses that are been achieved by the authors to be presented in future research.

Finally, related to the physiological variables generated by the Eye-Tracking, an important variable is the pupil diameter. Regarding to the investigation of the participants pupil diameter, the slope generated by the values of a participant pupil diameter during some time, i.e. the time spent during the evaluation of some graph or table, is also investigated in order to identify patterns of behavior.

Acknowledgments. This study was partially sponsored by the Coordination for the Improvements of Higher Education Personnel – Brazil (CAPES) and the Brazilian Research Council (CNPq) for which the authors are most grateful.

References

1. Keeney, R.L., Raiffa, H.: Decisions with Multiple Objectives: Preferences, and Value Tradeoffs. Wiley, New York (1976)
2. Belton, V., Stewart, T.: Multiple Criteria Decision Analysis. Kluwer Academic Publishers, Dordrecht (2002)
3. Greco, S., Ehrgott, M., Figueira, J.R. (eds.): Multiple Criteria Decision Analysis. ISORMS, vol. 233. Springer, New York (2016). https://doi.org/10.1007/978-1-4939-3094-4
4. Zamarrón-Mieza, I., Yepes, V., Moreno-Jiménez, J.M.: A systematic review of application of multi-criteria decision analysis for aging-dam management. J. Clean. Prod. **147**, 217–230 (2017)

5. Weber, M., Borcherding, K.: Behavioral influences on weight judgments in multi-attribute decision making. Eur. J. Oper. Res. **67**, 1–12 (1993)

6. de Almeida, A.T., de Almeida, J., Costa, A.P.C.S., De Almeida-Filho, A.T.: A new method for elicitation of criteria weights in additive models: flexible and interactive tradeoff. Eur. J. Oper. Res. **250**, 179–191 (2016)

7. Linkov, I., Cormier, S., Gold, J., Satterstrom, F.K.: Bridges, T: Using our brains to develop better policy. Risk Anal. Int. J. **32**(3), 374–380 (2012)

8. Loewenstein, G., Rick, S., Cohen, J.D.: Neuroeconomics. Ann. Rev. Psychol. **59**, 647–672 (2008)

9. Glimcher, P.W., Rustichini, A.: Neuroeconomics: the consilience of brain and decision. Science **5695**, 447–452 (2004)

10. Goucher-Lambert, K., Moss, J., Cagan, J.: Inside the mind: using neuroimaging to understand moral product preference judgments involving sustainability. J. Mech. Des. **139**(4), 041–103 (2017)

11. Khushaba, R.N., Wise, C., Kodagoda, S., Louviere, J., Kahn, B.E., Townsend, C.: Consumer neuroscience: assessing the brain response to marketing stimuli using electroencephalogram (EEG) and eye tracking. Expert Syst. Appl. **40**(9), 3803–3812 (2013)

12. Riedl, R., Davis, F.D., Hevner, A.R.: Towards a NeuroIS research methodology: intensifying the discussion on methods, tools, and measurement. J. Assoc. Inf. Syst. **15**(10), i–xxxv (2014). https://doi.org/10.17705/1jais.00377. Special Issue

13. Dimoka, A., Pavlou, P.A.; Davis, F.D.: Neuro-IS: the potential of cognitive neuroscience for information systems research. In: Proceedings of the 28th International Conference on Information Systems, pp. 1–20 (2007)

14. Hunt, L.T., Dolan, R.J., Behrens, T.E.: Hierarchical competitions subserving multi-attribute choice. Nat. Neurosci. **17**(11), 1613 (2014)

15. Nermend, K.: The implementation of cognitive neuroscience techniques for fatigue evaluation in participants of the decision-making process. In: Nermend, K., Łatuszyńska, M. (eds.) Neuroeconomic and Behavioral Aspects of Decision Making. SPBE, pp. 329–339. Springer, Cham (2017). https://doi.org/10.1007/978-3-319-62938-4_21

16. Trepel, C., Fox, C.R., Poldrack, R.A.: Prospect theory on the brain? Toward a cognitive neuroscience of decision under risk. Cogn. Brain. Res. **23**(1), 34–50 (2005)

17. Barberis, N., Xiong, W.: What drives the disposition effect? An analysis of a long-standing preference-based explanation. J. Financ. **64**(2), 751–784 (2009)

18. Roselli, L.R.P., de Almeida, A.T., Frej, E.A.: Decision neuroscience for improving data visualization of decision support in the FITradeoff method. Oper Res Int J **19**, 933–953 (2019). https://doi.org/10.1007/s12351-018-00445-1

19. Roselli, L.R.P., Frej, E.A., de Almeida, A.T.: Neuroscience experiment for graphical visualization in the FITradeoff decision support system. In: Chen, Y., Kersten, G., Vetschera, R., Xu, H. (eds.) GDN 2018. LNBIP, vol. 315, pp. 56–69. Springer, Cham (2018). https://doi.org/10.1007/978-3-319-92874-6_5

20. de Almeida, A.T., Roselli, L.R.P.: Visualization for decision support in FITradeoff method: exploring its evaluation with cognitive neuroscience. In: Linden, I., Liu, S., Colot, C. (eds.) ICDSST 2017. LNBIP, vol. 282, pp. 61–73. Springer, Cham (2017). https://doi.org/10.1007/978-3-319-57487-5_5

21. Klimesch, W., Schack, B., Sauseng, P.: The functional significance of theta and upper alpha oscillations. Exp. Psychol. **52**(2), 99–108 (2005)

22. Holm, A., Lukander, K., Korpela, J., Sallinen, M., Müller, K.M.I.: Estimating brain load from the EEG. Sci. World J. **9**, 639–651 (2009)

23. Macdonald, J.S.P., Mathan, S., Yeung, N.: Trial-by-trial variations in subjective attentional state are reflected in ongoing prestimulus EEG alpha oscillations. Front. Psychol. **2**, 82 (2011)

24. de Loof, E., et al.: Preparing for hard times: scalp and intracranial physiological signatures of proactive cognitive control. Psychophysiology **56**, 10 (2019)
25. Hines, W.W., Montgomery, D.C.: Probability and Statistics in Engineering and Management Science. Wiley, New York (1990)
26. Frej, E.A., de Almeida, A.T., Costa, A.P.C.S.: Using data visualization for ranking alternatives with partial information and interactive tradeoff elicitation. Oper. Res. Int. J. **2019**, 1–23 (2019)

The Multi-Actor Multi-Criteria Analysis (MAMCA): New Software and New Visualizations

He Huang[(✉)] [ID], Philippe Lebeau [ID], and Cathy Macharis [ID]

Vrije Universiteit Brussel, Pleinlaan 2, 1050 Brussel, Belgium
{he.huang,philippe.lebeau,cathy.macharis}@vub.be

Abstract. The Multi-Actor Multi-Criteria Analysis has been a successful methodology to integrate multiple stakeholders in the decision-making process. Because MAMCA evaluates different alternatives based on the objectives of the stakeholders, decision-makers can increase the support for the alternative they will choose. Still, the application of the methodology can be complex to popularize this approach. The MAMCA software was therefore published in order to facilitate the use of the methodology. The development of that tool offers also new opportunities. Currently, the goal is to extend the MAMCA software as a mass participation tool, hence maximizing participation involvement.

In order to facilitate the application of the methodology, the new MAMCA software was published. This contribution highlights how the MAMCA methodology was integrated into the software and how the data is being visualized. We focus on enhancing the concept of "Participation" in the development. A new data structure has been developed and an easier user interface makes the tool more accessible. An easy-understand evaluation method is integrated into the software. The interaction experience between participants is improved. Overall, the new MAMCA software is aimed to have a better performance in workshop settings.

Keywords: MAMCA · MCDM · Data visualization · Human-computer interaction

1 Introduction

Several types of operations research methods have been developed to help decision-makers evaluate transport projects. A common method to do this is Multiple-criteria decision analysis (MCDA) or Multiple-criteria decision-making (MCDM), ranking or sorting different alternatives based on at least two criteria [8]. MCDM has become more and more popular as it allows to evaluate different kinds of criteria (and not only economical ones). However, in practical transport cases, more than just one individual or group of individuals, called stakeholders, are involved, which can significantly influence or be influenced by the result of

© Springer Nature Switzerland AG 2020
J. M. Moreno-Jiménez et al. (Eds.): ICDSST 2020, LNBIP 384, pp. 43–56, 2020.
https://doi.org/10.1007/978-3-030-46224-6_4

the decision [10]. Crucial is thus to incorporate different points of view from several stakeholders into such an analysis. As the result, it can reveal the preferences of different stakeholders, hence allowing easier and clearer decision-making.

MAMCA, an extension of traditional MCDM methods, was proposed for transport project evaluations [18]. During the decision-making process, different stakeholders are taken into account. The concept of stakeholder is involved at the early stage of the evaluation, which leads to a better understanding of the objectives for different stakeholders. MAMCA successfully reflects the preferences of every individual stakeholder and expresses their concerns. It has been applied in various domains, especially in the field of mobility and logistics [1]. MAMCA was used in different scenarios such as evaluating transport policy measures [4] and transport technologies [19]. It has also proven itself as a useful methodology in transport-related decision making [2].

To facilitate the application of the MAMCA methodology, a web tool was developed, called MAMCA software [12]. Since 2016, the MAMCA software has helped decision-makers in different sectors to gain a better understanding of the MAMCA methodology and support them with decision-making. However, as time goes by, the limits of the original MAMCA software were exposed, mainly in the form of the difficulty of extending functions and outdated programming technology. Thus, new software is required to be developed to help MAMCA adapt to fast-paced technology changes, and capable of the situation that massive stakeholders can participate in the evaluation.

In this paper, we will first introduce the MAMCA methodology in Sect. 2. Section 3 presents the new MAMCA software and its distinct features. Finally, we will discuss the future directions made possible by the new MAMCA software in Sect. 4.

In order to present the features and illustrate visualizations of the software, a didactic last-mile case in the supply chain will be taken as an example.

1.1   Supply Chain Management Case Study

The case study entitled "The last-mile in the supply chain" is a fictive case study, but corresponds to real dilemma situations regarding home deliveries. It is aimed to gain insight into the extent to which different alternatives for the last mile of a supply chain for home deliveries contribute to the interests of the different stakeholder groups involved. As the stakeholder groups hold different priorities into different criteria, a multi-actor view is needed to show the different points of view of the stakeholder group. The list of alternatives and the criteria of the stakeholder groups are shown in Tables 1 and 2.

2   MAMCA Methodology

The steps of a classic MCDM process include the problem statement, alternatives and criteria definition, alternatives screening, scores determination, scores analysis, and drawing of conclusions [22]. Unlike classical MCDM methods, MAMCA

Table 1. The alternatives in the supply chain management case

Alternative name	Alternative description
Electric vehicles	Only electric vehicles are authorized to access the city center
Mobile depot & Cargo bikes	Free parkings are foreseen for trucks that split their final deliveries with cargobikes
Lockers delivered at night	Places are booked for companies in strategic areas in the city for lockers. They are delivered at night only
Crowdsourced deliveries	Online customers can choose to be delivered from a crowdsourced service
Business as usual	–

Table 2. The criteria of stakeholder groups in the supply chain management case

Stakeholder group	Citizens	Local authorities	Logistics service providers	Receivers	Shippers
Criteria	Road safety	Quality of life	Viability of investment	Low costs for receiving goods	Low cost deliveries
	Air quality	Network optimization	Profitable operations	Convenient high quality deliveries	High level service
	Urban accessibility	Social political acceptance	High level service	Attractive living environment	Positive impact on society
	Attractive urban environment	Positive business climate	Positive impact on society	Green concerns	Successful pick-ups
	Low noise nuisance		Employee satisfaction		

takes stakeholder analysis to identify stakeholder groups after defining alternatives. Each stakeholder group can have different criteria tree [17]. In Fig. 1, the overall methodology of MAMCA is shown.

In the first step, the potential alternatives to solve the problems are defined. The decision-makers need to identify and classify the alternatives in terms of different scenarios, policy measures and so on. In the second step, stakeholder analysis is taken to identify the stakeholders/stakeholder groups. It is a crucial step in MAMCA as for each stakeholder (group) there is a different criteria tree. An in-depth understanding of each stakeholder group is needed. Next, criteria and the corresponding weights are chosen and defined for each stakeholder group. One or more indicators for each criterion need to be constructed in step four. The indicators can be used to measure each alternative, providing the scale for the judgment.

In step 5, the overall analysis is taken within stakeholder groups. Any MCDM methods can be used to assess the alternatives. The Group decision support methods (GDSM) are well suited in this step such as Preference ranking organi-

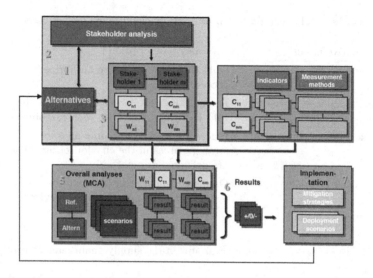

Fig. 1. The methodology of MAMCA [16]

zation method for enrichment evaluation (PROMETHEE) [3], analytic hierarchy process (AHP) [11]. There is no conflict between stakeholder groups and groups. The final evaluations and results of every stakeholder group will only be confronted at the end of the analysis.

The results of the analysis are presented in step 6. Additionally, a sensitivity analysis can be performed to check the robustness of the results. For each stakeholder (group), the multi-criteria analysis reveals their respective criteria and favored solutions, while the multi-actor multi-criteria analysis indicates the comparison of the different points of view of every stakeholder (group), which supports the decision-maker in making the final decision. Eventually, the actual implementation of the decision chosen is taken. The information collected from the previous steps helps the decision-maker to define the implementation paths.

3   The New MAMCA Software

To fulfill the need of MAMCA assessment with an interaction interface, MAMCA software was developed. However, the studies on MCDM increased every year, the innovated methodologies emerge and evolve fast [20]. The original version of MAMCA cannot integrate more MCDM methods because of the limitation of extendibility. In the workshop, it took time to introduce the MAMCA methodology and the MCDM method will be used in the evaluation. An efficient and simple MAMCA procedure is sought to speed up the workshop. Additionally, the higher capacity number of stakeholders for analysis is asked, to maximize the participation involvement. In order to make the evaluation within a stakeholder group with a large number of participants feasible, extending the MAMCA software as a mass participation tool is needed. By doing this, it is possible to get more opinions from a large stakeholder group like citizens.

Thus, a new version of MAMCA software with high extensibility has been developed to integrate new information technologies and visualizations[1]. It is written in the software stack of MongoDB, Node.js, Express, React (MERN) [24].

3.1 The Evaluation Steps and Visualizations

The new MAMCA software follows the evaluation structures of MAMCA methodology. In a MAMCA project assessment, the software divides it into 6 steps, which include alternatives identification, stakeholder group identification, criteria definition, criteria weight allocation, alternative evaluation, discussion and results.

After creating a MAMCA project, the project manager is able to define new alternatives, as well as modify and remove them. After defining alternatives, stakeholder groups are identified. Each stakeholder group is described according to the objectives they have regarding the alternatives. These objectives are the criteria used to evaluate the impact of scenarios on stakeholders' support. With these three first steps, the project manager has designed the architecture of the MAMCA projects. Data are then collected in the next steps to run the analysis.

In the fourth step, each criterion is weighted. The project manager or the stakeholders can manually allocate weights. Still, other allocation methods are proposed in the software. The stakeholders can choose the pairwise comparison, that they indicate their preference intensities for pairs of criteria.

Stakeholders can also use Direct Rating [7]. All criteria will be rated on a 100-point scale. The most important criterion will be given by the highest number. All other criteria are then rated in comparison to the most important one. The rated scores will be normalized. Suppose there is a set of criteria in one stakeholder group, calling $F = \{f_1, f_2, ..., f_m\}$. $W = \{W_1, W_2, ..., W_m\}$ is the set of given priority scores for the criteria, and $w = \{w_1, w_2, w_m\}$ is the normalised criteria weights set. The final weight of criterion k will be:

$$w_k = \frac{W_k}{\sum_{j=1}^{m} W_j} \tag{1}$$

In the fifth step, the stakeholders should evaluate the alternatives based on their criteria. Currently, two additional methods are available: AHP developed by Saaty and Simple Multiattribute Rating Technique (SMART) [6]. If AHP is chosen, pairwise comparison is conducted between alternatives.

If SMART is chosen, the preferences of the alternatives can be rated on a 10-point scale. Suppose one stakeholder has to evaluate ta finite set of alternative $A = \{a_1, a_2, ...a_n\}$ The performance score of P_i of alternative a_i will be calculated by means of weighted sums [9]:

$$P_i = \frac{\sum_{j=1}^{m} p_{ij} w_j}{10} \tag{2}$$

---

[1] For more information, please visit: https://mamca.vub.be/.

Overall result

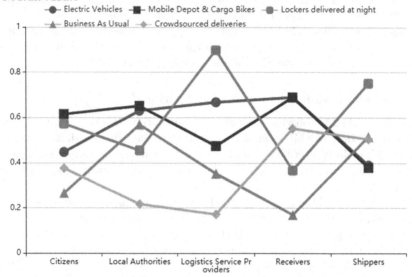

Fig. 2. The multi-actor analysis

Where p_{ij} is the performance score of alternative a_i on the criterion f_j, w_j is the weight of criterion f_j. The final performance score is divided by 10 in order to keep the score ranges from 0 to 1.

Once all stakeholders in one stakeholder group finished evaluating, the final performance score in the stakeholder group will be calculated in arithmetic mean. Say there are h stakeholders in a stakeholder group $X = \{X_1, X_2, X_3..., X_h\}$. The set of final scores F is thus:

$$F = \{F_i = \frac{\sum_{k=0}^{h} P_{ik}}{h}; i = 1, ..., n\} \tag{3}$$

Finally, after evaluation, the results are visualized. The new version distinct itself from the previous one, introducing lines with different marker symbols. This allows the lines to be more easily distinguished from one another, as well as to offer greater accessibility for black-and-white prints or color-blind readers. The Multi-Actor view as shown in Fig. 2 represents the final scores on different alternatives for each stakeholder group. The lines stand for the alternatives. It is easy to see that different stakeholder groups have different preferred alternatives. This chart represents the value of the MAMCA: it depicts clearly the support of each stakeholder for different solutions.

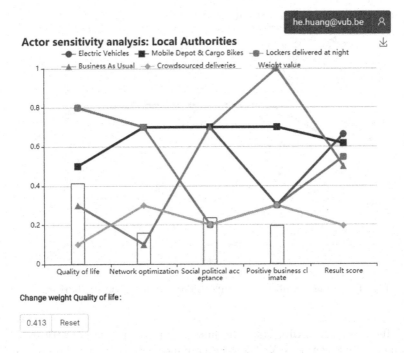

Fig. 3. Actor sensitivity analysis

The sensitivity analysis is integrated into the evaluation and weight chart. As shown in Fig. 3, The project manager is able to change the weights of criteria in any stakeholder group, hence allowing to check the robustness of the results. As shown in the figure, by clicking the button top-right corner, the project manager can check the weights allocation and evaluation results from different stakeholders in the stakeholder group "Local Authorities".

If there is more than one stakeholder in the stakeholder group, the box plot of the weights' difference can be shown when the project manager wants to check the average result of one stakeholder group. As shown in Fig. 4, the box plot of each weight indicates the difference of the weights allocation from different stakeholders. This visualization is especially beneficial when there is a large number of stakeholders in one stakeholder group. This allows the project manager to know if stakeholders are more controversial about the importance of some criteria while having an agreement on other criteria. For example, in Fig. 4, it can be seen that there is bigger deviance in the weight allocation of criterion "Quality of life", and a less deviance in the weight allocation of the criterion "Network optimization".

3.2 New Features in the Software

Besides the change of software stack, other major changes were made.

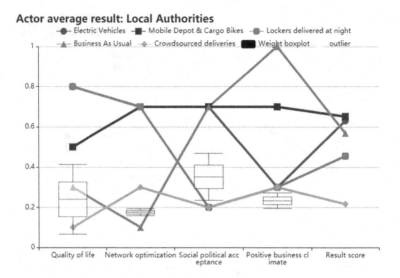

Fig. 4. Average result of the stakeholder group "Local Authorities"

High Effective Technologies. The first major change of the software is the replacement of web services. Web services are means to exchange data and information over the network. By building the web services, the frontend of the software and backend can be separated. The web services will communicate from the frontend and the backend of the software. The web services can be built based on two styles, the previous version of MAMCA software relying on Simple Object Access Protocol (SOAP). However, the other style, Representational State Transfer Protocol (REST), which was defined later, has a better throughput and response time. It has the definite advantage over the SOAP style [13].

Another major change of technology is the programming language. Java and PHP are used in the previous version of MAMCA software, which is robust and secure. Oppositely, the new MAMCA software is written in JavaScript, both frontend and backend. With the help of JavaScript, it is possible to make MAMCA a single-page application, that is, a web interface composed of individual components which can be reloaded independently [21]. So there will be no need for reload of the entire page, which can save more resources for the software. The data transaction between the frontend and backend is through JavaScript Object Notation (JSON). As a lightweight data carrier, it is human-readable and efficient. [15] The final result is that the new software has less response time than the previous one. To test the performance of the new software, a controlled trial was taken between new MAMCA software and the previous version of MAMCA software. The same project was chosen, and pairwise comparison was taken in the same stakeholder group "Local Authorities" to weigh the criteria. Network traffic was captured during the weighing, resources loaded and response time was recorded as shown in Table 3. The previous MAMCA software sent 49 requests to the server and loaded 1.2 MB resources in total. Oppositely, the new MAMCA software only sent 2 requests to the server and loaded 1.9 KB resources also with much faster response time.

Table 3. The performance of two versions of MAMCA software

	Previous MAMCA software	New MAMCA software
Request sent	49	2
Resources (kB)	1200	1.9
Response time (ms)	2816	42

New Database Structure. In the previous version of MAMCA software, the relational database is used. MySQL is used as the database management system. In the new MAMCA software, MongoDB is chosen, which is a NoSQL database. It is a document-oriented database, and the data is stored in JSON-like documents. It is more flexible than the SQL as it allows the different structure or fields. Talking about the performance of the database, MongoDB has higher reading and writing speed than the conventional SQL [14]. Furthermore, as a document-oriented database, the data of one project saved in MongoDB is not distributed in different database tables anymore, which is easier to collect and analyze for further data analysis.

3.3 Enhanced "Participation" Concept

It is necessary to get an idea of the needs and objectives of the stakeholders, that's the reason to develop the MAMCA software. The new MAMCA software is easier to involve more stakeholders in the decision-making process. And in this software, it can have an easier, faster way to evaluate and better comprehension. We did this, by the integration of the SMART method which is a very straight forward way to evaluate alternatives. The new participation system also improves the interaction experience between participants.

The Integration of SMART Method. It was observed during MAMCA workshops that the evaluators most often spent a lot of time to understand the theory of the MCDM method. Also, it was time-consuming when they did the pairwise comparison if there are many criteria. Thirdly, for many it was still perceived as a black box. That's why SMART is integrated in the software. As the oldest, simplest and most used MCDM method, the reason to apply this method into software is that stakeholders will be able to understand how their input is used to calculate preference scores, which is more unlikely in PROMETHEE and AHP. In contrast to AHP, there is no issue of stakeholders having to perform lots of pairwise comparisons. Another advantage of SMART is that the overall performance scores can be meaningfully interpreted, instead of being a dimensionless index that is only meaningful in comparison to other scores.

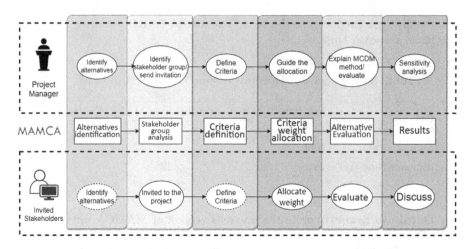

Fig. 5. The new participation system in the MAMCA software

Comparing to AHP, SMART sacrifices accuracy and sensitivity for its simplicity. Because of the subjective nature of technique, SMART is not consistent in contrast to the pairwise comparison. It is not suggested to use SMART method to make the final decision but a way to get insight into the objectives with different alternatives in a short time [23]. With SMART, the participants can save more time to comprehend the meaning of the performance score, and understand the importance of the presence of other stakeholders in the group: as shown in (2), it is easy to know the different weight allocation on criteria and the different preference on alternatives from other stakeholders will affect the final score of one alternative.

Easier Interaction Between Participants. The project manager and invited stakeholders can have a better experience in communication and comparison in the new software thanks to the new participation system. It helps facilitate the MAMCA evaluation, especially in a workshop. The Fig. 5 shows how the new participation system works.

After identifying the stakeholder groups in a MAMCA project, the project manager can invite stakeholders to the project through the email invitation. As the dashed circles in the Fig. 5 indicates, it is optional that the project manager and stakeholders can identify alternatives together. The stakeholders can also define the criteria of their groups with guidance from the project manager.

The project manager can coordinate the works of evaluators. For example, normally the weight allocation of criteria is more subjective than the evaluation of alternatives. The stakeholders can allocate weights based on their priority. Though when they evaluate the alternatives, they may need help from the experts. They can discuss the consensual performance scores of alternatives in the stakeholder group. The project manager can put the scores they discussed in the evaluation table. After that, the stakeholders are able to use the same performance scores from the project manager with one click of a button.

During the evaluation, the project manager is free to check the weight allocations and alternative evaluations from stakeholders. Also, as mentioned before, after the evaluation, the participants can check the average stakeholder group result. Stakeholders in the same stakeholder group are able to check the result of others. The project manager can do the sensitivity analysis, in order to reach the consensus among all. The new software expresses the differences of MAMCA from the other MCDM methodology: it searches the win-win solutions by taking the different points of view from stakeholders' accounts. The new software can help stakeholders understand the impact on each other.

4   Discussion and Future Work

The motivation for developing the new software is to make the MAMCA methodology more understandable and more accessible for the participants in the project. The software is refined for ease of use and reliability and is especially suitable for the evaluation in workshops. The integration of the SMART method allows participants to understand the evaluation steps, hence being more transparent.

Because of the characteristic of the development stack, the new MAMCA software is easy to extend functions, which means there can be more features to integrate into the future works.

Improvement on the Concept of "Participation". The first refinement for further work is to improve the concept of "Participation". Currently, the stakeholders can discuss the weight differences of the criteria, compare the performance scores they give to alternatives. In the end, the concept should be finished in a closed loop: a consensus-seeking mechanism is thus needed. Doan and De Smet developed an alternative weight sensitivity analysis based on linear programming (MILP) [5]. It can be applied in the MAMCA methodology to offer a consensus between different stakeholders by taking the inverse optimization point of view.

Integration of Other MCDM Methods. As any MCDM method can be used in the MAMCA methodology, especially the GDSM-method as they are able to cope with the stakeholder concept [17], other MCDM methods such as PROMETHEE can be integrated into the software thanks to easy extendibility of the software. By increasing the available methods the users have more freedom to choose suitable methods. For example, evaluators can use PROMETHEE as they provide different preference functions which suitable for different scenarios, or they can choose AHP because of its consistency.

Development for Mass Participation. Because of the flexibility and high performance of the new database in the software, it is prepared for mass participant involvement analysis. A stakeholder group such as citizens is able to include massive amounts of participants with different behaviors and preferences. Subgroups within one stakeholder group can be clustered based on their evaluation or preferences. A model will be designed to analyze and classify this large amount of data.

5   Conclusion

In this paper, the new MAMCA software was introduced to better support the decision-making process of the stakeholders. As the new interaction tool for MAMCA methodology, it follows the evaluation structures of the methodology with a simple and clear user interface. It is aimed to have a better performance in workshop settings. The SMART method is integrated to make the participants focus on understanding the meaning of their scoring instead of spending time to comprehend the theory of the MCDM method. The software enhances the concept of participation during the evaluation. Besides the representative result visualizations, sensitivity analysis and box plots of weight allocations within stakeholder groups are developed. The participants can have a better understanding of the influence of their behaviors and preferences.

The MAMCA software is designed as a tool to understand and analyze the role and input of stakeholders in strategic processes. It can be seen as a transition tool as participants learn to look at the decision problem in a new and more empathetic way. The uniqueness of MAMCA lies in the multi-actor evaluation, as stakeholders learn to see how other stakeholders might have other goals and criteria. In the evaluation process, the stakeholder is aware of the presence of the other stakeholders. There is a learning loop for the stakeholders. The stakeholder can have a better understanding of each other's position, which makes a stakeholder group more prone to search common solutions, to reach the consensus. The idea is that the habits of one individual should be altered, however not in an imposed way, but rather in a voluntary way. In addition to this, we should be aware that individual behavior is not happening on an island. In the end, the MAMCA software is not a tool to make the decision for the participants, but a tool to help them to understand and analyze the role and input of themselves in strategic processes.

References

1. Baudry, G., Macharis, C.: Decision-Making for Sustainable Transport and Mobility, Multi-Actor Multi-Criteria Analysis (2018). https://doi.org/10.4337/9781788111805
2. Baudry, G., Macharis, C., Vallée, T.: A range based multi-actor multicriteria analysis to incorporate uncertainty in stakeholder based evaluation processes (2014)
3. Brans, J.P., Vincke, P., Mareschal, B.: How to select and how to rank projects: the promethee method. Eur. J. Oper. Res. **24**(2), 228–238 (1986)

4. Crals, E., Keppens, M., Macharis, C., Ramboer, R., Vereeck, L., Vleugels, I.: Too many questions and too few options. Results of research into tradable mobility rights. Traffic Spec. **110**, 15–18 (2004)
5. Doan, N.A.V., De Smet, Y.: An alternative weight sensitivity analysis for PROMETHEE II rankings. Omega **80**, 166–174 (2018)
6. Edwards, W.: How to use multiattribute utility measurement for social decision-making. IEEE Trans. Syst. Man Cybern. **7**(5), 326–340 (1977)
7. Edwards, W., von Winterfeldt, D.: Decision Analysis and Behavioral Research, vol. 604, pp. 6–8. Cambridge University Press, Cambridge (1986)
8. Figueira, J.É., Greco, S., Ehrogott, M.: Multiple Criteria Decision Analysis: State of the Art Surveys. ISORMS, vol. 78. Springer, New York (2005). https://doi.org/10.1007/b100605
9. Fishburn, P.: Additive utilities with incomplete product set: applications to priorities and (1967)
10. Freeman, R.E.: Strategic Management: A Stakeholder Approach. Cambridge University Press, Cambridge (2010)
11. Golden, B.L., Wasil, E.A., Harker, P.T.: The Analytic Hierarchy Process. Applications and Studies. Springer, Heidelberg (1989). https://doi.org/10.1007/978-3-642-50244-6
12. Hadavi, S., Macharis, C., Van Raemdonck, K.: The multi-actor multi-criteria analysis (MAMCA) tool: methodological adaptations and visualizations. In: Żak, J., Hadas, Y., Rossi, R. (eds.) EWGT/EURO-2016. AISC, vol. 572, pp. 39–53. Springer, Cham (2018). https://doi.org/10.1007/978-3-319-57105-8_2
13. Kumari, S., Rath, S.K.: Performance comparison of soap and rest based web services for enterprise application integration. In: 2015 International Conference on Advances in Computing, Communications and Informatics (ICACCI), pp. 1656–1660. IEEE (2015)
14. Li, Y., Manoharan, S.: A performance comparison of SQL and NOSQL databases. In: 2013 IEEE Pacific Rim Conference on Communications, Computers and Signal Processing (PACRIM), pp. 15–19. IEEE (2013)
15. Lin, B., Chen, Y., Chen, X., Yu, Y.: Comparison between JSON and XML in applications based on AJAX. In: 2012 International Conference on Computer Science and Service System, pp. 1174–1177. IEEE (2012)
16. Macharis, C.: The importance of stakeholder analysis in freight transport (2005)
17. Macharis, C., De Witte, A., Ampe, J.: The multi-actor, multi-criteria analysis methodology (MAMCA) for the evaluation of transport projects: theory and practice. J. Adv. Transp. **43**(2), 183–202 (2009)
18. Macharis, C., Turcksin, L., Lebeau, K.: Multi actor multi criteria analysis (MAMCA) as a tool to support sustainable decisions: state of use. Decis. Support Syst. **54**(1), 610–620 (2012)
19. Macharis, C., Verbeke, A., De Brucker, K.: The strategic evaluation of new technologies through multicriteria analysis: the advisors case. Res. Transp. Econ. **8**, 443–462 (2004)
20. Mardani, A., Jusoh, A., Nor, K., Khalifah, Z., Zakwan, N., Valipour, A.: Multiple criteria decision-making techniques and their applications-a review of the literature from 2000 to 2014. Econ. Res.-Ekonomska Istraživanja **28**(1), 516–571 (2015)
21. Mesbah, A., Van Deursen, A.: Migrating multi-page web applications to single-page AJAX interfaces. In: 11th European Conference on Software Maintenance and Reengineering (CSMR 2007), pp. 181–190. IEEE (2007)
22. Nijkamp, P., Rietveld, P., Voogd, H.: Multicriteria Evaluation in Physical Planning, vol. 185. Elsevier (2013)

23. Patel, M.R., Vashi, M.P., Bhatt, B.V.: Smart-multi-criteria decision-making technique for use in planning activities. New Horisons in Civil Engineering (NHCE 2017), At Surat, Gujarat, India (2017)
24. Subramanian, V.: Pro MERN Stack: Full Stack Web App Development with Mongo. Express, React, and Node. Springer, Heidelberg (2017). https://doi.org/10.1007/978-1-4842-2653-7

Complexity Clustering of BPMN Models: Initial Experiments with the K-means Algorithm

Chrysa Fotoglou ⓘ, George Tsakalidis ⓘ, Kostas Vergidis$^{(\boxtimes)}$ ⓘ,
and Alexander Chatzigeorgiou ⓘ

Department of Applied Informatics, University of Macedonia, Thessaloniki, Greece
{chfotoglou,giorgos.tsakalidis,kvergidis,achat}@uom.edu.gr

Abstract. This paper introduces a method to assess the complexity of process models by utilizing a cluster analysis technique. The presented method aims to facilitate multi-criteria decision making and process objective management, through the combination of specific quality indicators. This is achieved by leveraging established complexity metrics from literature, and combining three complementary ones (i.e. NOAJS, CFC and CNC) to a single weighted measure, offering an integrated scheme for evaluating complexity. K-means clustering algorithm is implemented on 87 eligible models, out of a repository of 1000 models, and classifies them to corollary clusters that correspond to complexity levels. By assigning weighted impact on specific complexity metrics -an action that leads to the production of threshold values- cluster centroids can fluctuate, thus produce customized model categorizations. The paper demonstrates the application of the proposed method on existing business process models from relevant literature. The assessment of their complexity is performed by comparing the weighted sum of each model to the defined thresholds and proves to be a straightforward and efficient procedure.

Keywords: Business intelligence · Business process complexity · Data mining · Cluster analysis · Multi-criteria decision making · BPMN · K-means

1 Introduction

Business process management (BPM) is regarded as a set of methods aimed at supporting the design, analysis and optimization of processes [1]. This research field has emerged as an area of high interest among scholars, professionals and practitioners and inferably provides organizations with competitive superiority [2]. A core aspect of BPM is the utilization of modeling techniques to represent business processes (BPs) in an efficient and understandable way that facilitates the application of redesign initiatives. Process models are used to illustrate the main internal elements of BPs, including activities, sequence flow, dataflow and actors involved, as well as their relationships [3].

In today's rapidly evolving environment, organizations are facing the challenge of handling larger and increasingly complex processes. An important aspect of process models, that affects their capability of being analyzed, transformed and optimized, is their complexity [4]. This important quality characteristic, is often used interchangeably

© Springer Nature Switzerland AG 2020
J. M. Moreno-Jiménez et al. (Eds.): ICDSST 2020, LNBIP 384, pp. 57–69, 2020.
https://doi.org/10.1007/978-3-030-46224-6_5

with the terms understandability, modifiability and maintainability [5], as increased complexity often results in low comprehensiveness by process operators and higher error probability. Additionally, it is a common occurrence that low model quality and high complexity hinder the success of redesign initiatives [6]. Therefore, complexity assessment of BP models is regarded essential to evaluate a model's capability for modification and redesign. The latter motivated the authors to develop an assessment method that categorizes models, represented in the Business Process Model and Notation (BPMN) standard, regarding their complexity. The proposed approach incorporates cluster analysis on an eligible subset of 1000 process models [7] and extracts specific threshold values for the efficient categorization of future process models. The evaluation of BPMN complexity is achieved through the integration of a series of metrics to a single weighted measure. The main benefit of this approach is that it provides the capability of customization, depending on the purpose and priorities of the modeler.

2 Related Work

Measuring the complexity of BPs is a task that can be examined from many different standpoints [8]. Nevertheless, the quantification of the notion of complexity is almost always performed with the employment of defined metrics. BP complexity metrics proposed in literature, mostly constitute indicators of understandability, maintainability and error-proneness of a process model ([8, 9]) and at the same time their majority are adaptations of software complexity metrics ([9, 10]). Drawing an analogy between software and BPs, Cardoso et al. [8] propose the Number of Activities metric (NOA), which calculates activity complexity, and was inspired by lines-of-code (LOC) metric [12]. Considering the control elements of process models, the Number of Activities and Control-flow elements (NOAC) metric is introduced for well-structured models, along with the Number of Activities, Joints and Splits (NOAJS) metric for unstructured ones. Albeit these complexity metrics are useful and simple to calculate, it is highly important to complement other forms of complexity.

In [13], Cardoso introduces the Control-Flow Complexity (CFC) metric, which is inspired from McCabe's cyclomatic complexity [14]. This metric evaluates the complexity of XOR-split, OR-split, and AND-split constructs and aims to measure the impact of control-flow elements on the perceived complexity of a process. Another established approach [15] draws from network analysis and graph theory, presenting various structural metrics and dividing them into the categories *size*, *density*, *partitionability*, *connector interplay*, *cyclicity*, and *concurrency*. Apart from the NOA metric for the calculation of a process's size, Mendling also defines Diameter (diam) for process models as the length of the longest path from a start to an end node [15]. Various density metrics adapted by the same author, provide information regarding the relation between arcs and nodes in a model. The Coefficient of Connectivity (CNC) metric, earlier proposed by Latva-Koivisto [16] with the purpose of measuring the degree of complexity of a critical network, is used in regard to process models to calculate the ratio of arcs to nodes. A similar approach is used for the Density (Δ) metric, which "refers to the number of arcs divided by the maximum number of arcs for the same nodes" [15]. Additionally, the Average Connector Degree (ACD) in a process model demonstrates the number of nodes a connector is in average connected to and Maximum Connector Degree (MCD) the maximum number of nodes for a connector [15]. A high value of density metrics is an indicator of complexity and low understandability of a model [17]. Likewise, Kluza and Nalepa [18] proposed the Durfee Square Metric (DSM) and Perfect Square Metric (PSM) specifically constructed for BPMN models, aiming to include, not only the number of elements in a process model, but also their variety in measuring its complexity.

It becomes apparent that a model's complexity cannot be directly determined by only one type of metric [19]. Mendling implies that metrics should be interpreted in relation to other metrics, by pointing out several limitations, e.g., that models with more activities can be more understandable or that the density metric should consider the size of a model so as to be sufficient [15]. Overall, only a few attempts of combining complexity metrics ([20, 21]) or defining thresholds for model classification ([22–24]) appear in literature. Emphasizing this need for well-defined thresholds of complexity measures, Mendling et al. [24] utilize logistic regression and an adaptation of the ROC curves, and, Sánchez-González et al. [25] empirically evaluate certain structural metrics, including the CFC and ACD metrics. In Yahya et al. approach [26], threshold values are extracted for a vast amount of complexity metrics from literature, associating them with comprehensibility and modifiability of BP models. An evaluation mechanism proposed by Tsakalidis et al. [27] also leverages on widely used complexity metrics from literature. The proposed mechanism regards complexity as a major obstacle for the transformation of BP to a Directed Acyclic Graph (DAG) and evaluates it with the support of threshold values for selected metrics. These thresholds, if exceeded, reveal a model's need for normalization to reduce complexity and support the transformation of the model to a DAG. What is also evident in literature is that despite the substantial amount of research on complexity metrics and their definition, there is a lack of systematic approaches that: (a) combine metrics for the evaluation of quality and complexity, and, (b) define reference or threshold values for these metrics. Considering the above, the method proposed in this study offers a holistic approach and aims to address complexity assessment issues in a comprehensive way, that enables the recognition of appropriate reference or threshold values.

3 Complexity Assessment Methodology

The first step towards complexity assessment is the selection of suitable metrics. The three aspects of a model's complexity that have the strongest presence in research are activity complexity, control flow complexity and structural complexity [10]. Evidently, the three aspects combined constitute a comprehensive approach on complexity measurement, covering the most important elements of a model. Following the review on related research regarding complexity measurement, the authors select three representative complexity metrics (NOAJS, CFC and CNC) which mainly focus on size, control-flow and structuredness of a process model, respectively. The selection of metrics is based on the fact that they are established measures, their calculation is straightforward and have been extensively used in research (e.g., [19, 22, 28]). The selected metrics are defined in a generic way that covers most modeling standards and are defined by simple mathematical operations; hence, their calculation is a straightforward procedure.

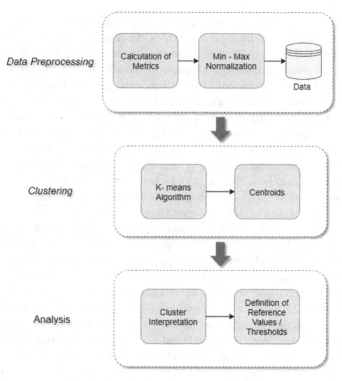

Fig. 1. Complexity assessment methodology

The Number of Activities, Joints and Splits (NOAJS) activity complexity metric stands as a representation of a process size, by simply counting the number of activities in a process, along with the number of split and join constructs. As a result, high values of this metric reveal a complex model, since complexity undeniably rises when the size of a process increases. Frequently, though, size metrics function, mainly collaboratively, as a point of reference for comparison purposes with other metrics [13]. Control-Flow Complexity (CFC) is regarded as a fundamental metric to evaluate a process model's complexity in relation to control flow. Since its introduction, it has been widely studied and empirically validated in research ([12, 14, 23]). The metric itself aims to quantify the cognitive effort of understanding the multiple states of a process after the occurrence of each type of split. The CFC metric should not be used in isolation to effectively evaluate the overall complexity of a BP; its combination with other complexity metrics is considered more efficient. Coefficient of Network Connectivity (CNC) exploits the notion of connectivity between elements to quantify structural complexity. High values of these metrics reveal a dense model, which indicates increased complexity and error probability [15]. The CNC of a process model is a simple measure to calculate and comprehend and it is defined as the ratio of arcs to nodes.

The selected metrics were introduced in a generic way that is applicable to many modeling languages. The BPMN standard comprises of a variety of elements, often absent in other modeling types, and, therefore, several problems may occur. For instance, the variety of elements present in a model might confuse as to what should be considered a node or an arc. To address this issue, the calculation of the metrics on BPMN models is conducted under the following assumptions: (a) all gateways (exclusive, inclusive and parallel) count as joins or splits, depending on their type, (b) all activities, gateways (exclusive, inclusive and parallel) and events (end, start and intermediate) count as nodes, (c) all sequence flow arrows count as arcs and (d) message flow arrows do not count as arcs.

Figure 1 displays an overview of the methodology followed to extract reference values for the assessment of complexity. First, a data preprocessing step is required to develop the dataset that will be used in the analysis. Next, the clustering phase follows, where the authors employ a data mining technique for the purpose of grouping data based on similarity, namely cluster analysis. The algorithm used in the current research is K-means, a popular method that groups datapoints by calculating centroid values for a pre-determined number of clusters. Last, by analyzing the resulting clusters, i.e., through their respective centroid values, reference values for the selected complexity metrics derive, facilitating the identification of complex models.

3.1 Data Preprocessing

In order to analyze and evaluate the notion of complexity for BPMN models, a repository of BPMN models is required. The SOA-based BP Database, comprised of 1000 BPs modeled in BPMN, constitutes a valuable source of models for the evaluation of our methodology. However, most of the processes (530) contained in this database do not represent real BPs, but only examples of theoretical ones hence before the extraction of the data, the preprocessing of the database is considered necessary. The steps followed to obtain the final dataset used in this work are shown in Fig. 2.

Some of the processes appeared to include errors that prohibited their examination (46). From the remaining processes, 155 of them were not complete, meaning they incorporate tasks that do not correspond to actual activities. The complete and actual processes in the database are described in a variety of languages, with the majority being

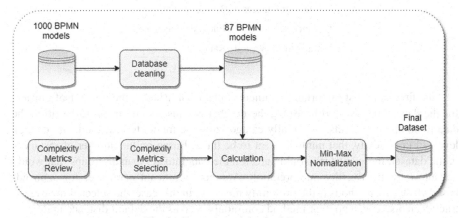

Fig. 2. Formation of the final dataset

in English (212). Spanish, French, German and Dutch are some of the languages used in the modeling of the remaining (67). In total, 212 complete processes in English and 14 translated processes were selected for further examination. A thorough analysis of the database led to the selection of 87 (8.7%) BPs, aligned with the requirements set (i.e. actual, complete BPs, containing no errors and labeled in English). These processes were examined and assigned an appropriate title to convey their context, since the database only identified the processes with an index number. Following the selection of the metrics and the establishment of the process pool to be used in the analysis, the calculation of each metric was performed according to the assumptions set on the previous section for each of the 87 valid BPMN models in the given database.

For the final dataset to be subjected to further analysis, the adjustment of the calculated metric values was necessary, given that the values of each metric are in different scale. Especially, in clustering analysis, data normalization is crucial to compare similarities between features based on certain distance measures. The method utilized, namely Min-Max normalization, is a simple, yet efficient way of rescaling data, widely used as a preprocessing step in data mining [30].

3.2 Clustering

The development of the evaluation methods proposed in this paper, is accomplished through cluster analysis and, more specifically, using the K-means algorithm [31]. Cluster analysis is, essentially, the grouping of data points in categories called clusters, based on similarity. The selected algorithm is one of the most popular ones for unsupervised learning problems and is found to deliver reliable results [32]. Several parameters require definition for the application of the K-means algorithm, as shown in Table 1.

Table 1. Input parameters

Parameter	Input
Algorithm	Simple K-means
Number of clusters	3
Distance measure	Euclidean distance
Initialization method	Random

The first and most important parameter to be established is the number of clusters for the data to be grouped in. Since the number of clusters are meant to partition the data to complexity levels, essentially categorizing the model to low, moderate or high levels of complexity, that number is set to be three. Based on the number of instances in the dataset, more clusters would partition the data into very small groups that would not allow for trustworthy interpretation. Another significant parameter for centroid-based clustering methods is the proximity measure. In this case, the selected measure is Euclidean distance, a popular method commonly used as the default distance metric for many cluster analysis tools [33]. Lastly, the K-means algorithm requires an initialization method to assign the first three centroid values. Random initialization, during which the initial centroids are randomly placed in the Euclidean space, is chosen, since during the experiments no need for a more sophisticated initialization method was revealed.

3.3 Analysis

The main concept behind the implementation of K-means on the metric values (features) of a set of BPMN models is related to the notion of categorizing similar instances together. Considering that each cluster, i.e., category, contains models of the same complexity, it is possible to assign a complexity level to each cluster. The centroid values of each cluster, acting as its representatives, reveal the complexity of the models in that cluster. For the purpose of this analysis, in conjunction with the definition of the metrics used for quantifying complexity, the authors deduce that higher centroid values correspond to high complexity, moderate centroid values correspond to moderate complexity and low values to low complexity. Essentially, the centroid values act as reference points for future categorization. The assessment of a new BPMN model's complexity is performed by discovering to which cluster, i.e., complexity category, it belongs.

4   Evaluation of BPMN Models

Drawing from related research on complexity metrics [20], a method to combine NOAJS, CFC and CNC into a weighted sum is introduced. Our method constitutes a way to evaluate complexity using information extracted from all three of the selected metrics combined to a single measure. Ultimately, it is a way to assign a priority number to each metric, expressed by a specific weight. The existing method we utilized for the combination of the metrics into a weighted sum is introduced in [20] and it is deemed appropriate, since it is a scalable and computationally efficient way of reducing the dimensionality of the problem concerned. In particular, the weighting method in its generic form can be applied to any number of metric values; in our study the selected metrics are 3. Assigning priority to metrics offers the potential for researchers to customize the assessment method, according to their perspective on complexity evaluation. Essentially, the weighting of the features prior to the implementation of the algorithm allows for the creation of a novel model. Additionally, the described approach enables the extraction of specific threshold values for each complexity level. By combining the metrics to one feature, the centroids can be represented by a one-dimensional vector, which in turn means that they can be depicted on a line. Therefore, the mean point of the intervals between centroids per two acts as a separator for the clusters.

Two scenarios are analyzed to demonstrate the above. The first scenario is a simple combination of the metrics, with equal weighting for each one. Thresholds are extracted, based on this method that facilitates the separation of the complexity levels. The second scenario considers the weighting parameter, by lowering the priority of a metric, to demonstrate the impact this option has on the threshold values. There are multiple options when it comes to assigning priority to metrics that can serve the purpose of each researcher. In this example, the priority of the control-flow metric, i.e., CFC, is significantly higher, while the priority of the size metric, i.e., NOAJS, is lowered relatively to the other metrics.

4.1 Scenario 1: Equal Priority

The first scenario to be examined concerns equal weighting for the three metrics. This entails that each metric has a priority of 1 ($p_{1,2,3} = 1$), which in turn means each metric weights 0.333. Following the defined procedure, the weighted sum measure is calculated for all instances. At this point, a cluster analysis on the one-dimensional array of the weighted sum feature is possible. The results are illustrated in Fig. 3. It is evident that the instances and their distance from the centroids can be represented on a straight line, including values from 0 to 1. It appears that only a small percentage (8%) of instances are grouped in Cluster 3, while the majority belongs to Clusters 1 and 2.

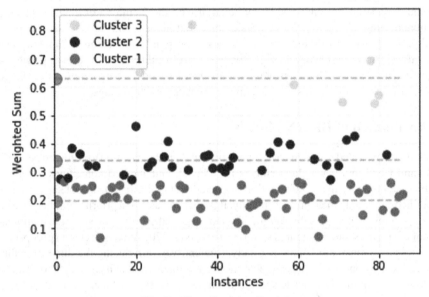

Fig. 3. Clustering visualization

The centroid values provided by the analysis are displayed in Table 2.

Table 2. Centroid values – Scenario 1

Centroid	Value
Centroid 1	0.197
Centroid 2	0.339
Centroid 3	0.630

Essentially, these values constitute reference points on a straight line. In this way, comparing the distance between centroids and assigning an instance to a cluster becomes a straightforward task, since the defined radius of a cluster is easily distinguished. Given the weighted sum of a future process model, the latter will be assigned to the cluster having a centroid value closest to its weighted sum. Figure 4 illustrates the clusters and their centroid values on a straight line. Based on the definition of the K-means algorithm and the parameters used in this case, the cluster assignment is determined by the Euclidean distance. This means that the points that separate the clusters, and act as threshold values, can be decided through a simple calculation of the means of the formed intervals between the cluster centroids. Pertaining to the defined assessment methodology, it is possible to draw conclusions regarding the type of models each cluster represents, based on the centroid values. Thus, we deduce that Cluster 1 contains low complexity models, Cluster 2 moderate complexity models and Cluster 3 high complexity ones. The assessment of a new BPMN model's complexity is possible with a straightforward comparison to the threshold values.

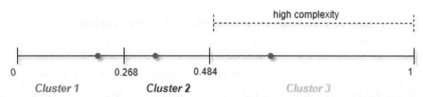

Fig. 4. Visualization of centroids and thresholds – Scenario 1 (Cluster 1: Low Complexity, Cluster 2: Moderate Complexity, Cluster 3: High Complexity)

4.2 Scenario 2: High CFC and Low NOAJS Priority

The weighted sum approach offers the option of assigning priority to the selected metrics and developing a customized model for assessing complexity. Let's assume that the purpose of an analysis dictates that a model's size should not influence the evaluation of complexity, while the control flow complexity of a model, i.e., the amount of gateway splits, should be mainly considered. This approach allows for the assignment of priorities accordingly, e.g., $p_1 = 1$, $p_2 = 10$, $p_3 = 5$, which means that NOAJS has a very low priority, CFC has a high priority, while CNC a moderate one.

Following the same process as in scenario 1, a cluster analysis is performed on the new weighted sum feature. Next, the instances are grouped to the three clusters again using the K-means algorithm. Different centroid values derive from the clustering, resulting in different thresholds for the complexity categorization. The new thresholds

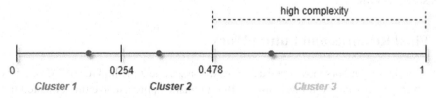

Fig. 5. Visualization of centroids and thresholds – Scenario 2 (Cluster 1: Low Complexity, Cluster 2: Moderate Complexity, Cluster 3: High Complexity)

and centroids are visible in Fig. 5. The difference between scenario 1 thresholds and scenario 2 is minor given the specific dataset, but still it is expected to have an impact on the categorization of new models.

4.3   Examples from Literature

This section features the application of the method developed to examples deriving from relevant literature. The assessment of their complexity is performed by comparing the weighted sum of each new BPMN model to the defined thresholds. The first step of the assessment involves the calculation of the metric values and the standardization process, followed by the calculation of the weighted sum of each model and, ultimately, the complexity assessment.

Table 3. Complexity assessment results on BPMN example cases

Process	Scenario 1		Scenario 2	
	Weighted sum	Complexity	Weighted sum	Complexity
Evaluate quote process [34]	**0.434**	**Moderate**	**0.524**	**High**
Property valuation [35]	0.387	Moderate	0.372	Moderate
Healthcare scenario [36]	0.463	Moderate	0.453	Moderate
Medical assessment [37]	0.506	High	0.571	High
Loan application [38]	**0.236**	**Low**	**0.275**	**Moderate**
Bank account opening [39]	0.547	High	0.649	High
Baking workflow process [40]	0.320	Moderate	0.319	Moderate
Emergency ward of a hospital [41]	0.498	High	0.559	High
Auction [42]	0.544	High	0.641	High
Admission process [43]	0.378	Moderate	0.388	Moderate

In Table 3 the assessment results for the 10 process models examined are presented. In most cases the two scenarios indicate the same complexity level. However, it can be observed, that between the two scenarios the complexity levels of two processes shift from moderate to high and from low to moderate respectively. This outcome relates to the increased priority granted to the CFC metric, which in turn raises the weighted sum measure accordingly.

5   Final Remarks and Future Work

In this paper, the authors introduced a complexity assessment method for BPMN models that leverages metrics from literature to offer a holistic approach. We implemented the K-means algorithm on an eligible subset of a 1000 BPMN models' database, to group them to clusters that correspond to complexity levels according to the selected metrics.

The combination of the metrics to a single weighted sum measure offers the capability of customizing the evaluation method based on a single objective. By adjusting the priorities of the metrics and analyzing the impact of that modification on their perceived complexity, the identification of problematic areas is supported. Additionally, this approach limits the dimensions of the data from three to one, facilitating the definition of exact threshold values that separate the complexity levels. The defined levels of complexity, i.e., clusters, are separated by the means of the intervals formed between two adjacent centroids and, in essence, act as thresholds. Therefore, the complexity classification of new BPMN models can be simple and straightforward.

However, there are limitations regarding the application of the assessment method. The development of the method involves the utilization of a data mining technique, particularly the implementation of the K-means algorithm for clustering. Such learning algorithms produce a model based on instances of data; hence they are largely dependent on the dataset provided. The results of the algorithms and the threshold values defined, are influenced in a significant degree by the nature of the dataset. An additional limitation of this research regards the number of divisions used to describe complexity. This number, i.e., three complexity levels, corresponds to the number of clusters to be formed by the algorithm and is an input parameter provided by the authors, according to the purpose of the analysis. Further partitioning of the complexity levels, for example considering four or five complexity levels, would yield different results.

Future work on complexity assessment will be directed towards tackling these limitations. The application of the proposed method on a larger dataset, containing BP models representative of industry standards, could provide a more accurate assessment of complexity. The same method could apply to a different set of metrics, pertinent to the context of each analysis and the purpose of the researcher.

References

1. Weske, M.: Business Process Management: Concepts, Architectures. Springer, Heidelberg (2012). https://doi.org/10.1007/978-3-642-28616-2
2. Tsakalidis, G., Vergidis, K., Delias, P., Vlachopoulou, M.: A conceptual business process entity with lifecycle and compliance alignment. In: Freitas, P.S.A., Dargam, F., Moreno, J.M. (eds.) EmC-ICDSST 2019. LNBIP, vol. 348, pp. 70–82. Springer, Cham (2019). https://doi.org/10.1007/978-3-030-18819-1_6
3. Lindsay, A., Downs, D., Lunn, K.: Business processes—attempts to find a definition. Inf. Softw. Technol. **45**, 1015–1019 (2003)
4. La Rosa, M., Wohed, P., Mendling, J., Ter Hofstede, A.H., Reijers, H.A., van der Aalst, W.M.: Managing process model complexity via abstract syntax modifications. IEEE Trans. Industr. Inf. **7**, 614–629 (2011)
5. de Oca, I.M.-M., Snoeck, M., Reijers, H.A., Rodríguez-Morffi, A.: A systematic literature review of studies on business process modeling quality. Inf. Softw. Technol. **58**, 187–205 (2015)
6. Kock, N., Verville, J., Danesh-Pajou, A., DeLuca, D.: Communication flow orientation in business process modeling and its effect on redesign success: results from a field study. Decis. Support Syst. **46**, 562–575 (2009)
7. BPOSCTeam SOA-Based Business Process Database. https://sites.google.com/site/bposcteam2015/ressources. Accessed 10 Nov 2019

8. Cardoso, J., Mendling, J., Neumann, G., Reijers, H.A.: A discourse on complexity of process models. In: Eder, J., Dustdar, S. (eds.) BPM 2006. LNCS, vol. 4103, pp. 117–128. Springer, Heidelberg (2006). https://doi.org/10.1007/11837862_13

9. Muketha, G.: A survey of business processes complexity metrics. Inf. Technol. J. **9**, 1336–1344 (2010)

10. Polančič, G., Cegnar, B.: Complexity metrics for process models – a systematic literature review. Comput. Stand. Interfaces. **51**, 104–117 (2017). https://doi.org/10.1016/j.csi.2016.12.003

11. Sánchez González, L., García Rubio, F., Ruiz González, F., Piattini Velthuis, M.: Measurement in business processes: a systematic review. Bus. Process Manage. J. **16**, 114–134 (2010)

12. Jones, C.: Programming Productivity. McGraw-Hill College, New York (1986)

13. Cardoso, J.: Business process control-flow complexity: metric, evaluation, and validation. Int. J. Web Serv. Res. (IJWSR) **5**, 49–76 (2008)

14. McCabe, T.J.: A complexity measure. IEEE Trans. Software Eng. **4**, 308–320 (1976)

15. Mendling, J.: Metrics for Process Models: Empirical Foundations of Verification, Error Prediction, and Guidelines for Correctness. Springer, Heidelberg (2008). https://doi.org/10.1007/978-3-540-89224-3

16. Latva-Koivisto, A.M.: Finding a complexity measure for business process models. Systems Analysis Laboratory, Helsinki University of Technology (2001)

17. Dumas, M., La Rosa, M., Mendling, J., Mäesalu, R., Reijers, H.A., Semenenko, N.: Understanding business process models: the costs and benefits of structuredness. In: Ralyté, J., Franch, X., Brinkkemper, S., Wrycza, S. (eds.) CAiSE 2012. LNCS, vol. 7328, pp. 31–46. Springer, Heidelberg (2012). https://doi.org/10.1007/978-3-642-31095-9_3

18. Kluza, K., Nalepa, G.J.: Proposal of square metrics for measuring business process model complexity. In: 2012 Federated Conference on Computer Science and Information Systems (FedCSIS), pp. 919–922. IEEE (2012)

19. Kluza, K., Nalepa, G.J., Lisiecki, J.: Square complexity metrics for business process models. In: Mach-Król, M., Pełech-Pilichowski, T. (eds.) Advances in Business ICT. AISC, vol. 257, pp. 89–107. Springer, Cham (2014). https://doi.org/10.1007/978-3-319-03677-9_6

20. Makni, L., Khlif, W., Zaaboub Haddar, N., Ben-Abdallah, H.: A tool for evaluationg the quality of business process models. In: INFORMATIK 2010–Business Process and Service Science–Proceedings of ISSS and BPSC (2010)

21. Khlif, W., Ben-Abdallah, H., Ben Ayed, N.E.: A methodology for the semantic and structural restructuring of BPMN models. Bus. Process Manage. J. **23**, 16–46 (2017)

22. Yahya, F., Boukadi, K., Ben-Abdallah, H., Maamar, Z.: A fuzzy logic-based approach for assessing the quality of business process models. In: ICSOFT, pp. 61–72 (2017)

23. Augusto, A., et al.: Automated discovery of process models from event logs: review and benchmark. IEEE Trans. Knowl. Data Eng. **31**, 686–705 (2018)

24. Mendling, J., Sánchez-González, L., García, F., La Rosa, M.: Thresholds for error probability measures of business process models. J. Syst. Softw. **85**, 1188–1197 (2012)

25. Sánchez-González, L., García, F., Ruiz, F., Mendling, J.: Quality indicators for business process models from a gateway complexity perspective. Inf. Softw. Technol. **54**, 1159–1174 (2012)

26. Yahya, F., Boukadi, K., Ben-Abdallah, H.: Improving the quality of business process models: lesson learned from the state of the art. Bus. Process Manage. J. (2018). https://doi.org/10.1108/BPMJ-11-2017-0327

27. Tsakalidis, G., Vergidis, K., Kougka, G., Gounaris, A.: Eligibility of BPMN models for business process redesign. Information. **10**, 225 (2019)

28. Oukharijane, J., Yahya, F., Boukadi, K., Abdallah, H.B.: Towards an approach for the evaluation of the quality of business process models. In: 2018 IEEE/ACS 15th International Conference on Computer Systems and Applications (AICCSA), pp. 1–8. IEEE (2018)

29. Rolón, E., Cardoso, J., García, F., Ruiz, F., Piattini, M.: Analysis and validation of control-flow complexity measures with BPMN process models. In: Halpin, T., et al. (eds.) BPMDS/EMMSAD -2009. LNBIP, vol. 29, pp. 58–70. Springer, Heidelberg (2009). https://doi.org/10.1007/978-3-642-01862-6_6

30. Al Shalabi, L., Shaaban, Z., Kasasbeh, B.: Data mining: a preprocessing engine. J. Comput. Sci. **2**, 735–739 (2006)

31. Lloyd, S.: Least squares quantization in PCM. IEEE Trans. Inf. Theory **28**, 129–137 (1982)

32. Jain, A.K.: Data clustering: 50 years beyond K-means. Pattern Recogn. Lett. **31**, 651–666 (2010)

33. Kassambara, A.: Practical Guide to Cluster Analysis in R: Unsupervised Machine Learning. In: STHDA (2017)

34. Kolar, J., Dockal, L., Pitner, T.: A dynamic approach to process design: a pattern for extending the flexibility of process models. In: Grabis, J., Kirikova, M., Zdravkovic, J., Stirna, J. (eds.) PoEM 2013. LNBIP, vol. 165, pp. 176–190. Springer, Heidelberg (2013). https://doi.org/10.1007/978-3-642-41641-5_13

35. Kannengiesser, U.: Can we engineer better process models? In: DS 58-1: Proceedings of ICED 09, the 17th International Conference on Engineering Design, Vol. 1, Design Processes, Palo Alto, CA, USA, 24–27 August 2009, pp. 527–538 (2009)

36. Knuplesch, D., Reichert, M., Fdhila, W., Rinderle-Ma, S.: On enabling compliance of cross-organizational business processes. In: Daniel, F., Wang, J., Weber, B. (eds.) BPM 2013. LNCS, vol. 8094, pp. 146–154. Springer, Heidelberg (2013). https://doi.org/10.1007/978-3-642-40176-3_12

37. Herbert, L.T., Sharp, R., Hansen, M.R.: Specification, Verification and Optimisation of Business Processes: A Unified Framework (2014)

38. Rogge-Solti, A., Weske, M.: Prediction of business process durations using non-Markovian stochastic Petri nets. Inf. Syst. **54**, 1–14 (2015)

39. Poizat, P., Salaün, G., Krishna, A.: Checking business process evolution. In: International Workshop on Formal Aspects of Component Software, Besançon, France, pp. 36–53 (2016)

40. Herbert, L., Hansen, Z.N.L., Jacobsen, P., Cunha, P.: Evolutionary optimization of production materials workflow processes. Procedia CIRP **25**, 53–60 (2014)

41. Mannhardt, F., de Leoni, M., Reijers, H.A., van der Aalst, W.M.P.: Data-driven process discovery - revealing conditional infrequent behavior from event logs. In: Dubois, E., Pohl, K. (eds.) CAiSE 2017. LNCS, vol. 10253, pp. 545–560. Springer, Cham (2017). https://doi.org/10.1007/978-3-319-59536-8_34

42. Fan, S., Zhimin, H., Storey, V.C., Zhao, J.L.: A process ontology based approach to easing semantic ambiguity in business process modeling. Data Knowl. Eng. **102**, 57–77 (2016)

43. Bukhsh, Z.A., van Sinderen, M., Sikkel, K., Quartel, D.A.: Understanding modeling requirements of unstructured business processes. In: ICE-B. pp. 17–27 (2017)

Case Studies and Applications

To Click or Not to Click? Deciding to Trust or Distrust Phishing Emails

Pierre-Emmanuel Arduin(✉)

Université Paris-Dauphine, PSL, DRM UMR CNRS 7088,
Place du Maréchal de Lattre de Tassigny, 75775 Paris Cedex 16, France
pierre-emmanuel.arduin@dauphine.psl.eu

Abstract. While the email traffic is growing around the world, such questions often arise to recipients: to click or not to click? Should I trust or should I distrust? When interacting with computers or digital arte-facts, individuals try to replicate interpersonal trust and distrust mechanisms in order to calibrate their trust. Such mechanisms rely on the ways individuals interpret and understand information.

Technical information systems security solutions may reduce external and technical threats; yet the academic literature as well as industrial professionals warn on the risks associated with insider threats, those coming from inside the organization and induced by legitimate users.

This article focuses on phishing emails as an unintentional insider threat. After a literature review on interpretation and knowledge man-agement, insider threats and security, trust and distrust, we present a methodology and experimental protocol used to conduct a study with 250 participants and understand the ways they interpret, decide to trust or to distrust phishing emails. In this article, we discuss the preliminary results of this study and outline future works and directions.

Keywords: Insider threats · Trust · Interpretation · Knowledge management · Decision-making

1 Introduction

Technical and externally centred Information Systems security solutions allow the prevention of intrusions [17], the detection of denial of service attacks [68], and the strengthening of firewalls [46]. Nevertheless, the academic literature as well as industrial professionals consider that a predominant threat is neither technical nor external, but human and inside the organization [29,54,64,66]. Such an insider threat may be intentional or non intentional, malicious or non malicious [5,35,65].

In fact, according to audit and advisory surveys such as [28], more than 33% of reported cyber-attacks between 2016 and 2018 used phishing, just behind those who used malware (36%). The proportion of insiders among threats increased from 46% in 2016 to 52% in 2018. According to cybersecurity ventures, employees should be trained to recognize and react to phishing emails [43].

© Springer Nature Switzerland AG 2020
J. M. Moreno-Jiménez et al. (Eds.): ICDSST 2020, LNBIP 384, pp. 73–85, 2020.
https://doi.org/10.1007/978-3-030-46224-6_6

More than 90% of successful attacks rely on phishing, *i.e.* emails leading their recipients to interpret and decide to trust them [21]. Recipients are invited, suggested or requested to click on a link, open a document or forward information to someone they should not.

For authors such as [2], security concerns and actual behaviour are disconnected due to the "lack of comprehension". It may be difficult for users to understand security warnings [14], as well as to identify ongoing attacks [20]. In this paper we argue that such understanding difficulties may be studied by focusing on trust and distrust elements users rely on when receiving an email. A logo, a date, a number or an email address, those elements and others are used to decide either an email may be trusted or not.

In the first section of this paper, background theory and assumptions are presented: First, the ways individuals interpret and understand information relying on the knowledge management literature; Second, insider threats and their different categories; Third, trust and distrust with a particular focus on individual psychological processes. In the second section of this paper, a study that involved 250 participants is presented: First, the methodology and experimental protocol; Second, a discussion of the preliminary results; Third, a presentation of future works. The overall purpose of this paper is to share observation statements, preliminary results, and future expectations on ways to prevent insider threats by identifying how we decide to trust or distrust phishing emails.

2   Background Theory and Assumptions

In this section, we first draw from the knowledge management literature to present the ways individuals interpret and understand information. Second, we discuss the importance of considering insider threats and their different categories. Third, we expose individual psychological processes leading to trust or distrust.

2.1   On Interpretation and Understanding

To describe the complex interpretation machinery, some authors talk about a "mental model" [23] or a "neural apparatus" [25], a place of chemical reactions that can be analysed. Others believe that interpretation involves above all the socio-individual [67], resulting from our history, a place for expressing a form of intellectual creativity specific to each person. We all act as interpretative agents, information processors interacting with the world that surrounds us through a filter. In fact, this indescribable filter through which we interact with the world may be called an *"interpretative framework"* [63].

Information is transmitted by talking, writing or acting during a *sense-giving* process. We collect data from this information by listening, reading or watching during a *sense-reading* process. *Sense-giving* and *sense-reading* processes are defined by [50] as follows: "Both the way we endow our own utterance with meaning and our attribution of meaning to the utterances of others are acts of tacit knowing. They represent sense-giving and sense-reading within the structure of tacit knowing" [50, p. 301]. When he studied the processes of *sense-giving* and *sense-reading*, [63] highlighted the idea that knowledge was the result of the interpretation by an individual of information.

Information is continuously created during sense-giving processes and interpreted during sense-reading processes. Knowledge can then be:

– *made explicit, i.e.* it has been made explicit by someone within a certain context, it is sense-given and socially constructed. Individuals, as well as computers are "information processing systems" [19, p. 9];
– *tacit, i.e.* it has been interpreted by someone within a certain context, it is sense-read and individually constructed. Relying on [49]: "We can know more than we can tell".

So that made explicit knowledge is tacit knowledge that has been made explicit by someone within a certain context. It is information source of tacit knowledge for someone else. It is "what we know and can tell" answering to [49] quoted above. Every piece of information can be seen as a piece of knowledge that has been made explicit by someone within a certain context and with their own intentions.

When a person P_1 structures his/her tacit knowledge and transmits it, he/she creates made explicit knowledge, *i.e.* information created from his/her tacit knowledge. A person P_2 perceiving this information and absorbing it, potentially creates new tacit knowledge for him/herself (see Fig. 1). Knowledge is the result of the interpretation by an individual of information. This interpretation is done through an interpretative framework that filters the data contained in the information and with the use of pre-existing tacit knowledge [63].

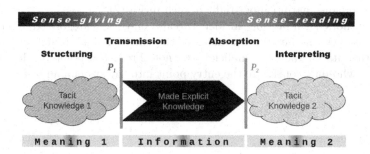

Fig. 1. Sense-giving and sense-reading: the ways we create, interpret and understand information

This interpretation leads to the creation of meaning that can vary from one individual to another: this is *meaning variance* [3, 4]. This question of meaning variance is central in organizations, notably for deciding whether an email may be trusted or not to prevent insider threats.

2.2 On Insider Threats

At the beginning of the 1990s, the literature on information systems security had already affirmed that there was "a gap between the use of modern technology and the understanding of the security implications inherent in its use" [35, p. 173]. The massive arrival of microcomputers was also accompanied by questions regarding the security of interconnected systems where computer science was previously mainframe oriented.

Indeed, the number of technological artefacts has exploded and this increase has gone hand in hand with the evolution of their various uses [9]. Yesterday, a terminal connected the user to the computer, while today entry points into the information system are multiple, universal, interconnected and increasingly discreet. Employee's social activity can be supported by social networks and their health maintained using connected watches.

The taxonomy of threats targeting the security of information systems proposed by [35] presented in Fig. 2 is disturbingly topical, with regard to the four dimensions that make up his angle of analysis: (1) sources, (2) perpetrators, (3) intent, and (4) consequences. It should be recognized that independent of the sources, perpetrators, and intent of a threat, the consequences remain the same: disclosure (of profitable information), modification or destruction (of crucial information), or denial of service (by hindering access to resources). These consequences are covered in the 2013 ISO/IEC 27001 standard: information security management, which defines information security management systems as ensuring (1) confidentiality, (2) integrity and (3) availability of information [22].

A business's firewall constitutes a protection against external threats, which appear on the left branch in Fig. 2. Authors such as [66] represent a part of the literature on information systems security that tends to pay attention to insider threats, more particularly those whose perpetrators are humans with the intention to cause harm (upper right branch in Fig. 2).

For authors such as [5], insider threats may be categorized along two dimensions: (1) whether the character of the threat is intentional or not, and (2) whether its character is malicious or not. From the point of view of the employee, who may constitute the entry point into the system, an insider threat can be:

Fig. 2. Taxonomy of IS security threats (inspired from [35])

1. *unintentional*: wrong actions taken by an inexperienced or negligent employee, or one manipulated by an attacker; for example, an inattentive click, input error, accidental deletion of sensitive data, etc. [59];
2. *intentional and non-malicious*: deliberate actions by an employee who derives a benefit but has no desire to cause harm; for example, deferring backups, choosing a weak password, leaving the door open during a sensitive discussion, etc. [15];
3. *intentional and malicious*: deliberate actions by an employee with a desire to cause harm; for example, divulging sensitive data, introducing malicious software into the computer system, etc. [58].

The study presented in this article focuses on the manipulation and social engineering techniques that exploit unintentional insider threats (category 1 above). Even though the attacker is outside the system and the organization, he makes an employee, a component of the system, unintentionally facilitate his/her infiltration: the latter has, for example, clicked on a link or even opened the door to a self-proclaimed delivery person with a self-proclaimed task. A social engineer is an attacker who targets a legitimate user from whom he/she obtains a direct (rights of access, harmful link visited, etc.) or indirect (vital information, relationship of trust, etc.) means to get into the system [42].

As new technological solutions are developed, the exploitation of hardware or software weaknesses becomes more and more difficult. Attackers are then turning toward another component of the system susceptible to attack: the human one. For authors as [56]: "Security is a process, not a product". For others such as [42, p. 14], breaching the human firewall is "easy", requiring no investment, except for occasional telephone calls and involves minimum risk. Every legitimate user constitutes thus an unintentional insider threat to the information system's security.

Individuals are not trained to be suspicious of others. Consequently, they constitute a strongest threat to the security of the information system insofar as any well-prepared individual can win their trust.

2.3 On Trust and Distrust

Even if trust is recognized as particularly important in security issues of computer networking environments [26], very little studies on information systems deal with both trust and security [51]. Some authors focus on end-users' trustworthiness [1,60] and others on trustworthy information systems' design [48,53,62].

In human and social sciences, authors as [37] and [47] consider that the psychological functionality of trust is to reduce the perceived uncertainty, *i.e.* the perceived risk in complex decision-making situations. Trust induces a mental reduction of the field of possibilities leading to take a decision without considering the outcome of each possible alternative [33].

Some authors consider concepts such as interpersonal trust and organizational trust, between respectively two or more people [13,24,57]. Others consider systemic trust, toward institutions or organizations [37], and trust in technologies [30,39].

An overall definition of trust seems to be lacking when tackling the literature. Relying on the taxonomy of [6], [44,45] defined trust as expectations: expectation of persistence of the natural and moral social orders, expectation of competence, and expectation of responsibility.

[51, p. 116] proposed an operational definition of trust as a "state of expectations resulting from a mental reduction of the field of possibilities". Such a definition appears to be consistent with the concept of distrust, which is a "confident negative expectation regarding another's conduct" [32, p. 439]. Distrust is often presented as relying on [36] and his suggestion that those who choose not to trust "must adopt another negative strategy to reduce complexity" [27, p. 24]. So, you trust when you have positive expectations, you distrust when you have negative expectations. Distrust should not be confused with mistrust, which is "either a former trust destroyed, or former trust healed" [61, p. 27] and is not considered in the study presented in this article.

[10, p. 7] went deeper when they stated that "the quantitative dimensions of trust are based on the quantitative dimensions of its cognitive constituents". These constituents are the beliefs on which we rely to trust, and they may explain the contents of our expectations. Examples related to trusted humans are: benevolence, integrity, morality, credibility, motives, abilities, expertise [38,40]. Examples related to trusted technologies are: dependability, reliability, predictability, failure rates, false alarms, transparency, safety, performance [16,18,34,55].

An appropriate distrust fosters protective attitudes [32] and reduces insider threats. Nevertheless, authors such as [51, p. 118] consider that "trust and distrust are alive, they increase or decrease depending on how expectations are met (or unmet [...])". Initial trust is notably based on information from third parties, reputation, first impressions, and personal characteristics such as the disposition to trust [41]. Then, from facts, understanding of the trustee's characteristics, predictability and limits notably [31,47], the trustor calibrates his/her trust [11,45]. Trust is adjusted, meaning expectations are adjusted.

3   Research Proposal and Experimental Protocol

In this section, we first present the methodology and experimental protocol we used to conduct a study with 250 participants and the ways they interpret, decide to trust or distrust phishing emails. Second, we discuss the preliminary results. Third, we outline future works and directions following this work-in-progress.

3.1   Description of the Study

The study has been conducted with 250 students of the Paris-Dauphine university. Half of them were Computer Science students and the other half Management Science students. Half of them were Bachelor students and the other half Master's degree students. Participants were given course credits for participating in the study. In the following we refer to the students involved in the study as the "participants". The average age of participants is 20.2 years.

A research engineer scheduled the presence of participants in a room with 10 computers and copy-protection walls. For the first waves of answers, a member of the research team was here to explain the purpose of the study and answer questions. Then the research engineer continued to manage the response room during one month in order to collect data and he was available to answer questions participants might have.

A short video presented the study to participants who were then given 20 emails, 8 of which were in English. They viewed each email one at a time on the computer screen and they were then asked to: (1) click on the areas leading them to trust it, (2) click on the areas leading them to distrust it, and (3) comment their choices in a general remarks field. Finally, participants should answer some profiling questions (age, academic level, etc.).

Participants arrived in the response room and gave informed consent to participate. Once installed, the researcher asked them if they had any questions. A short video gave them instructions:

This animation will present you the objective of this questionnaire, as well as the perspectives of the study. It will introduce the way you have to understand the questions in order to improve the usefulness of your answers for our investigation. Each question is composed of three parts:
1. *click on the areas of the email that make you think that it is official;*
2. *click on the areas of the email that make you think that it is fraudulent;*
3. *comment in a few words.*
The results of the first part will allow to identify elements of trust carried by the mail, whereas the results of the second part will allow to identify elements of distrust carried by the mail.
The free text box "comments" allows you every time to explain your feeling. When you are ready, click "start" below.

Then participants completed the task, as described above.

3.2 Discussion of the Preliminary Results

Data has been managed by the online reaction time experiments solution Qualtrics (see [7]) notably in order to produce heatmaps [8] such as shown in Fig. 3. General remarks fields have been manually tagged by the research team and when interpretation doubts were encountered, participants were contacted to clarify their meaning.

In this article, we consider the subset of 8 English emails in order to restrict the amount of data to process. Thus the collected data represents for each participant and each email, trust and distrust areas, meaning: $(250 \times 8) \times 2 = 4\,000$ images of emails with clicked areas. These images have been aggregated by the Qualtrics solution to $8 \times 2 = 16$ heatmaps: 8 trust-leading areas images and 8 distrust-leading areas images. Figure 3 shows three examples of trust-leading and distrust-leading areas in emails. As outlined in the next section, the research team is now analysing such images, notably to find invariant elements used as

trust or distrust givers by individuals, *i.e.* elements leading participants to decide
to trust or to distrust.

We also have to consider that each participant could explain their choices
for each email in a general remarks field. Such data represent $(250 \times 8) = 2\ 000$
open text fields explaining the choices. A preliminary analysis of the open text
fields related to the 8 English emails highlights that participants first focused on
elements leading them to distrust emails. Even if they were asked to select trust
and distrust areas in emails, only 7.6% of the participants mentioned both trust
and distrust elements in their written explanations, the rest of them focusing
only on distrust elements. This may be a bias of the study, probably induced
by the material of the study: phishing emails. The heatmaps generated by the
Qualtrics solution allow to partially bypass such a bias by analysing both trust-
leading and distrust-leading areas.

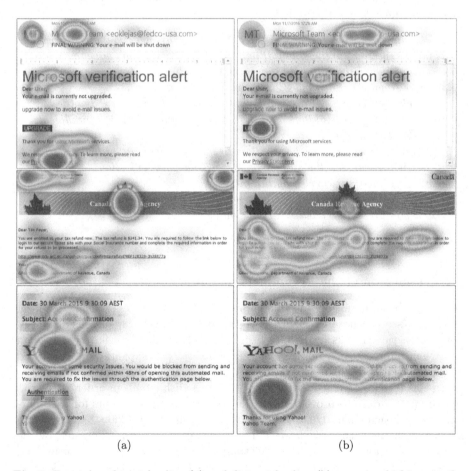

(a)   (b)

Fig. 3. Examples of trust-leading (a) and distrust-leading (b) areas in phishing emails

Participants who mentioned trust elements in their written explanation listed the presence of privacy concerns (6%) or logos (1.6%) in the emails. 44.1% of the participants mentioned the sender's address as a distrust element in their written comments. 25% of the participants mentioned the presentation of the email, *i.e.* typeface, structure, and spelling, as a distrust element, and 14% of them mentioned the pressure or emergency as a distrust element. Few of them (less than 2%) mentioned the presence of a link (particularly non-HTTPS) or an attachment to download, the occurrence of terms such as "secure", the absence of a human contact or personal data as distrust elements.

3.3 Presentation of Future Works

It is obvious that the results presented in this article are attached to the set of the study, meaning the 250 participants: students in higher education whose average age is 20.2 years. The purpose of the study is not to cover all the trust and distrust mechanisms that individuals put in place when interacting with computers or digital artefacts. The study presented in this article is still in progress and it aims to share observation statements, preliminary results, and future expectations on ways to prevent insider threats by identifying elements individuals rely on when deciding to trust or distrust phishing emails.

By understanding the ways individuals interpret, understand, trust or distrust an email, this study intents to prevent manipulation techniques hackers can use in order to influence individuals' decision to trust. Authors as [12, p. 92] stated that "in most of [the] studies no attempt was made to differentiate between the survey samples drawn from those who intentionally violate the procedures and policies and drawn from those who unintentionally violate them". The case of phishing emails is particularly interesting because existing behavioural countermeasures such as improving awareness, installing a rule-oriented organizational culture, deterrence or neutralization mechanisms show their limits on sloppiness and ignorance [12].

Currently, the research team is going deeper in the analysis of the collected data, working particularly on the overall set and not only the English emails. We aim to observe invariant elements used as trust or distrust givers by individuals. Such elements may be used by hackers as well as by security teams in organizations to adapt their formations and future actions.

As explained in Sect. 2.2 such a study focuses on unintentional insider threats notably caused by sloppiness or ignorance. In the future, we plan to tackle intentional insider threats, when individuals intentionally violate the information systems' security policy.

4 Conclusions and Perspectives

In this article, we focus on a particular threat for information systems' security: the unintentional and insider threat represented by individuals receiving phishing emails. Such individuals may facilitate the infiltration of an attacker despite themselves, by deciding to trust a phishing email.

In the first section, we presented the ways individuals interpret information relying on the knowledge management literature, the landscape of insider threats and their specificities, trust and distrust mechanisms involved in complex decision-making situations. In the second section, we presented a study conducted with 250 participants in order to highlight trust and distrust leading areas in emails and understand trust and distrust elements used by participants when receiving phishing emails.

The decision to trust, as well as manipulation techniques, were involved in decision-making situations well before the introduction of computers. In this article we studied the ways individuals interpret information to understand how they decide to trust or distrust phishing emails. Ethical issues should not be neglected in such a research, notably by considering the risk of dual-use of research results, as stated by [52]. The reader has to be aware that the results of this research may be used maliciously to mislead recipients.

The study presented in this article is a work-in-progress and the research team is now going deeper by analyzing the overall set of responses. In a near future it is planed to tackle another threat for information systems security: the intentional and insider threat represented by individuals deciding to intentionally violate the information systems' security policy. Such a study will be realized within industrial fields due to its potential managerial causes and implications.

References

1. Aberer, K., Despotovic, Z.: Managing trust in a peer-2-peer information system. In: Proceedings of the Tenth International Conference on Information and Knowledge Management, pp. 310–317. ACM (2001)
2. Anderson, B., Bjornn, D., Jenkins, J., Kirwan, B., Vance, A.: Improving security message adherence through improved comprehension: neural and behavioral insights. In: 2018 Americas Conference on Information Systems (AMCIS). AIS (2018)
3. Arduin, P.-E.: On the use of cognitive maps to identify meaning variance. In: Zaraté, P., Kersten, G.E., Hernández, J.E. (eds.) GDN 2014. LNBIP, vol. 180, pp. 73–80. Springer, Cham (2014). https://doi.org/10.1007/978-3-319-07179-4_8
4. Arduin, P.E.: On the measurement of cooperative compatibility to predict meaning variance. In: Proceedings of IEEE International Conference on Computer Supported Cooperative Work in Design (CSCWD), Calabria, Italy, 6–8 May, pp. 42–47 (2015)
5. Arduin, P.E.: Insider Threats. Wiley, New York (2018)
6. Barber, B.: The Logic and Limits of Trust. Rutgers University Press, New Brunswick (1983)
7. Barnhoorn, J.S., Haasnoot, E., Bocanegra, B.R., van Steenbergen, H.: QRTEngine: an easy solution for running online reaction time experiments using qualtrics. Behav. Res. Methods **47**(4), 918–929 (2015). https://doi.org/10.3758/s13428-014-0530-7

8. Bojko, A.A.: Informative or misleading? Heatmaps deconstructed. In: Jacko, J.A. (ed.) HCI 2009. LNCS, vol. 5610, pp. 30–39. Springer, Heidelberg (2009). https://doi.org/10.1007/978-3-642-02574-7_4

9. Canohoto, A., Dibb, S., Simkin, L., Quinn, L., Analogbei, M.: Preparing for the future - how managers perceive, interpret and assess the impact of digital technologies for business. In: Proceedings of the 48th Hawaii International Conference on System Sciences, Kauai, HI (2015)

10. Castelfranchi, C., Falcone, R.: Trust is much more than subjective probability: mental components and sources of trust. In: Proceedings of the 33th Hawaii International Conference on System Sciences, Piscataway, NJ (2000)

11. Costé, B., Ray, C., Coatrieux, G.: Trust assessment for the security of information systems. In: Pinaud, B., Guillet, F., Gandon, F., Largeron, C. (eds.) Advances in Knowledge Discovery and Management. SCI, vol. 834, pp. 159–181. Springer, Cham (2019). https://doi.org/10.1007/978-3-030-18129-1_8

12. Crossler, R.E., Johnston, A.C., Lowry, P.B., Hu, Q., Warkentin, M., Baskerville, R.: Future directions for behavioral information security research. Comput. Secur. **32**, 90–101 (2013)

13. Deutsch, M.: Trust and suspicion. J. Conflict Resolut. **2**(4), 265–279 (1958)

14. Felt, A.P., et al.: Improving SSL warnings: comprehension and adherence. In: Proceedings of the 33rd Annual ACM Conference on Human Factors in Computing Systems, pp. 2893–2902. ACM (2015)

15. Guo, K., Yuan, Y., Archer, N., Connely, C.: Understanding nonmalicious security violations in the workplace: a composite behavior model. J. Manag. Inf. Syst. **28**(2), 203–236 (2011)

16. Hancock, P.A., Billings, D.R., Schaefer, K.E., Chen, J.Y., De Visser, E.J., Parasuraman, R.: A meta-analysis of factors affecting trust in human-robot interaction. Hum. Factors **53**(5), 517–527 (2011)

17. Hansen, J.V., Lowry, P.B., Meservy, R.D., McDonald, D.M.: Genetic programming for prevention of cyberterrorism through dynamic and evolving intrusion detection. Decis. Support Syst. **43**(4), 1362–1374 (2007)

18. Hasselbring, W., Reussner, R.: Toward trustworthy software systems. Computer **39**(4), 91–92 (2006)

19. Hornung, B.: Constructing sociology from first order cybernetics: basic concepts for a sociocybernetic analysis of information society. In: Proceedings of the 4th Conference of Sociocybernetics, Corfu, Greece (2009)

20. Hu, Q., Dinev, T., Hart, P., Cooke, D.: Managing employee compliance with information security policies: the critical role of top management and organizational culture. Decis. Sci. **43**(4), 615–660 (2012)

21. Hurley, R.: The decision to trust. Harvard Bus. Rev. **84**, 55–62 (2006)

22. ISO/IEC: ISO/IEC 27001, information security management. Technical report (2013)

23. Jones, N., Ross, H., Lynam, T., Perez, P., Leitch, A.: Mental models: an interdisciplinary synthesis of theory and methods. Ecol. Soc. **16**(1), 46 (2011)

24. Kramer, R.M.: Trust and distrust in organizations: emerging perspectives, enduring questions. Annu. Rev. Psychol. **50**(1), 569–598 (1999)

25. Kuhn, T.: Reflections on my critics. In: Criticism and the Growth of Knowledge. Cambridge University Press (1970)

26. Lamsal, P.: Understanding trust and security. Department of Computer Science, University of Helsinki, Finland (2001)

27. Lane, C., Bachmann, R., Bachmann, L.: Trust Within and Between Organizations: Conceptual Issues and Empirical Applications. Oxford University Press, Oxford (1998)
28. Lavion, D.: PwC's global economic crime and fraud survey 2018. Technical report (2018)
29. Leach, J.: Improving user security behaviour. Comput. Secur. **22**(8), 685–692 (2003)
30. Lee, J.D., See, K.A.: Trust in automation: designing for appropriate reliance. Hum. Factors **46**(1), 50–80 (2004)
31. Lewicki, R.J., Bunker, B.B.: Developing and maintaining trust in work relationships. In: Trust in Organizations: Frontiers of Theory and Research, pp. 114–139 (1996)
32. Lewicki, R.J., Mc Allister, D.J., Bies, R.J.: Trust and distrust: new relationships and realities. Acad. Manag. Rev. **23**(3), 438–458 (1998)
33. Lewis, J.D., Weigert, A.: Trust as a social reality. Soc. Forces **63**(4), 967–985 (1985)
34. Li, X., Hess, T.J., Valacich, J.S.: Why do we trust new technology? A study of initial trust formation with organizational information systems. J. Strateg. Inf. Syst. **17**(1), 39–71 (2008)
35. Loch, K.D., Carr, H.H., Warkentin, M.E.: Threats to information systems: today's reality, yesterday's understanding. MIS Q. **16**, 173–186 (1992)
36. Luhmann, N.: Trust and Power. Wiley, Chichester (1979)
37. Luhmann, N.: Familiarity, confidence, trust: problems and alternatives. Trust: Making Breaking Coop. Relat. **6**, 94–107 (2000)
38. Mayer, R.C., Davis, J.H., Schoorman, F.D.: An integrative model of organizational trust. Acad. Manag. Rev. **20**(3), 709–734 (1995)
39. Mc Knight, D.H., Carter, M., Thatcher, J.B., Clay, P.F.: Trust in a specific technology: an investigation of its components and measures. ACM Trans. Manag. Inf. Syst. (TMIS) **2**(2), 12 (2011)
40. McKnight, D.H., Chervany, N.L.: Trust and distrust definitions: one bite at a time. In: Falcone, R., Singh, M., Tan, Y.-H. (eds.) Trust in Cyber-societies. LNCS (LNAI), vol. 2246, pp. 27–54. Springer, Heidelberg (2001). https://doi.org/10.1007/3-540-45547-7_3
41. McKnight, D.H., Chervany, N.L.: Handbook of Trust Research, pp. 29–51 (2006)
42. Mitnick, K., Simon, W.: The Art of Deception: Controlling the Human Element of Security. Wiley, New York (2003)
43. Morgan, S.: Cybercrime damages $ 6 trillion by 2021. Technical report (2016)
44. Muir, B.M.: Trust between humans and machines, and the design of decision aids. Int. J. Man Mach. Stud. **27**(5–6), 527–539 (1987)
45. Muir, B.M.: Trust in automation: part i. Theoretical issues in the study of trust and human intervention in automated systems. Ergonomics **37**(11), 1905–1922 (1994)
46. Ayuso, P.N., Gasca, R.M., Lefevre, L.: FT-FW: a cluster-based fault-tolerant architecture for stateful firewalls. Comput. Secur. **31**, 524–539 (2012)
47. Numan, J.: Knowledge-based systems as companions. Trust, human computer interaction and complex systems. Ph.D. thesis, Groningen, NL (1998)
48. Offor, P.I.: Managing risk in secure system: antecedents to system engineers' trust assumptions decisions. In: 2013 International Conference on Social Computing (SocialCom), pp. 478–485. IEEE (2013)
49. Polanyi, M.: Personal Knowledge: Towards a Post Critical Philosophy. Routledge, London (1958)
50. Polanyi, M.: Sense-giving and sense-reading. Philos.: J. Roy. Inst. Philos. **42**(162), 301–323 (1967)

51. Rajaonah, B.: A view of trust and information system security under the perspective of critical infrastructure protection. Ingénierie des Systèmes d'Information **22**(1), 109 (2017)
52. Rath, J., Ischi, M., Perkins, D.: Evolution of different dual-use concepts in international and national law and its implications on research ethics and governance. Sci. Eng. Ethics **20**(3), 769–790 (2014)
53. Ruotsalainen, P., Nykänen, P., Seppälä, A., Blobel, B.: Trust-based information system architecture for personal wellness. In: MIE, pp. 136–140 (2014)
54. Sasse, M.A., Brostoff, S., Weirich, D.: Transforming the 'weakest link'—a human/computer interaction approach to usable and effective security. BT Technol. J. **19**(3), 122–131 (2001)
55. Schaefer, K.E., Chen, J.Y., Szalma, J.L., Hancock, P.A.: A meta-analysis of factors influencing the development of trust in automation: implications for understanding autonomy in future systems. Hum. Factors **58**(3), 377–400 (2016)
56. Schneier, B.: The process of security. Inf. Secur. **3**(4), 32 (2000)
57. Schoorman, F.D., Mayer, R.C., Davis, J.H.: An integrative model of organizational trust: past, present, and future. Acad. Manag. Rev. **32**(2), 344–354 (2007)
58. Shropshire, J.: A canonical analysis of intentional information security breaches by insiders. Inf. Manag. Comput. Secur. **17**(4), 221–234 (2009)
59. Stanton, J., Stam, K., Mastrangelo, P., Jolton, J.: Analysis of end user security behaviors. Comput. Secur. **24**(2), 124–133 (2005)
60. Swamynathan, G., Zhao, B.Y., Almeroth, K.C.: Decoupling service and feedback trust in a peer-to-peer reputation system. In: Chen, G., Pan, Y., Guo, M., Lu, J. (eds.) ISPA 2005. LNCS, vol. 3759, pp. 82–90. Springer, Heidelberg (2005). https://doi.org/10.1007/11576259_10
61. Sztompka, P.: Trust: A Sociological Theory. Cambridge Cultural Social Studies. Cambridge University Press, Cambridge (1999)
62. Truong, N.B., Um, T.W., Lee, G.M.: A reputation and knowledge based trust service platform for trustworthy social internet of things. In: Innovations in Clouds, Internet and Networks (ICIN), Paris, France (2016)
63. Tsuchiya, S.: Improving knowledge creation ability through organizational learning. In: ISMICK 1993: Proceedings of the International Symposium on the Management of Industrial and Corporate Knowledge, pp. 87–95 (1993)
64. Vroom, C., Von Solms, R.: Towards information security behavioural compliance. Comput. Secur. **23**(3), 191–198 (2004)
65. Warkentin, M., Willison, R.: Behavioral and policy issues in information systems security: the insider threat. Eur. J. Inf. Syst. **18**(2), 101–105 (2009)
66. Willison, R., Warkentin, M.: Beyond deterrence: an expanded view of employee computer abuse. MIS Q. **37**(1), 1–20 (2013)
67. Yamakawa, Y., Naito, E.: From physical brain to social brain. In: Cognitive Maps. InTech (2010)
68. Zhi-Jun, W., Hai-Tao, Z., Ming-Hua, W., Bao-Song, P.: MSABMS-based approach of detecting LDoS attack. Comput. Secur. **31**(4), 402–417 (2012)

Management Information System for Police Facility Location

Ana Paula Henriques de Gusmão[1(✉)], Bruno Ferreira da Costa Borba[2],
and Thárcylla Rebecca Negreiros Clemente[2]

[1] Departamento de Engenharia de Produção, Universidade Federal de Sergipe (UFS) Campus
São Cristóvão, São Cristóvão, SE, Brazil
anapaulagusmao@cdsid.com
[2] Management Engineering CAA, Universidade Federal de Pernambuco, Av. Campina Grande,
s/n - Km 59 - Nova Caruaru, Caruaru, PE 55014-900, Brazil
brunoborba50@hotmail.com, thnegreiros@ymail.com

Abstract. It is well known to society that crime rates in Brazil have been greatly increasing in recent years. The Northeast region of Brazil is ranked second in the ranking of reported crimes in Brazil. The Northeast state of Pernambuco, in the last two years, has invested considerable sums in public safety and its crime rates have not fallen. In fact, several factors make it difficult to make decisions on prioritizing investments, such as the lack of a system to support management decisions. Considering that the location of police installations influences crime rates, specifically in police response time, this paper proposes a Management Information System (MIS) that suggests different location arrangements, with their respective performance indicators, to support decisions regarding the siting of police units. The MIS model for siting police facilities uses two methods, the first to find potential locations is called K-means, and the second to select optimal locations is called the Maximum Covering Location Problem (MCLP). The results obtained when using MIS in a Brazilian city, based on the occurrence of crimes, to identify police units, demonstrated the service performed better performance when compared to the current configuration of the units.

Keywords: Facility location problem · Public safety · Management Information System · K-means · MCLP

1 Introduction

In recent years, the growth of crime rates in Brazil and other Latin American countries has come into greater focus. This rapid growth in crime rates generates a sense of fear in the population, especially those living in large cities, including state capitals, where the indicators of violence indicate a sharp increase in crime rates [1]. In Pernambuco, while these rates have increased, at the same time, investments in public safety have also been significantly increased, including in three categories (Policing, Civil Defense and Information and Intelligence) which increased by 97.49% for Information and Intelligence

J. M. Moreno-Jiménez et al. (Eds.): ICDSST 2020, LNBIP 384, pp. 86–98, 2020.
https://doi.org/10.1007/978-3-030-46224-6_7

over the previous year [2]. Yet, even by investing more, there has been no reduction in crime rates.

In order to minimize these crime rates and maximize police resources, key executives and decision makers (DMs) have taken a keener and closer interest in using Operational Research (OR) techniques [3, 4]. Thus, one of the areas of study that has been very promising over time, with regard to public safety, is the optimal location of police units focused on increasing security and preserving order [4–7].

However, according to [8], the construction or purchase of new police facilities is typically an expensive and sensitive project over time. Before a project is implemented, an evaluation must be made as to whether the locations of police units are viable in terms of their operational and strategic goals. Therefore, OR is used to help DMs in this difficult task of choosing a location that will not only result in better performance vis-à-vis the current structure, but is promising in the future, even if environmental factors change or evolve [9].

Because of these facts, a DM faces multiple and complex difficult issues, as they must consider several factors in order to maximize the return on investment when siting a police facility. Considering the above, this paper proposes that a Management Information System (MIS) be developed to support managerial and operational issues related to siting police facilities, with a view to identifying alternative locations that maximize the return on investments based on the quality of service to be provided.

2 Theoretical Background

2.1 Covering Problems

The Maximum Covering Location problem (MCLP) model comes as a solution to the DM for facility location problems that need to cover a desired service distance. However, the DM cannot solely focus on covering a region that serves the largest number of customers, but must also consider the response time for providing the service to customers within that region of coverage [10].

Therefore, the service distance in most cases is linked to the response time associated with appropriately equipped police officers moving from the installation to the point of demand [11]. This means that there is no point in establishing a very long distance S that will cover more points of demand, as the response time for reaching the farthest point adversely affects the quality of service provided, by police stations, hospitals and fire brigades [8, 10, 12].

The mathematical structure of the MCLP model, which was proposed by [10] in 1974, is presented using the following equations.

$$\text{Max} \sum_{i\in I}^{n} a_i y_i \tag{1}$$

$$\text{subjetc to} \sum_{j\in N_i} x_j \geq y_i \ \forall i \in I \tag{2}$$

$$\sum_{j\in J} x_j = P \tag{3}$$

$$x_j \in \{0, 1\} \quad \forall j \in J \tag{4}$$

$$y_{ij} \in \{0, 1\} \quad \forall i \in I \tag{5}$$

Where:

I = the set of demand nodes;
J = the set of potential operational locations;
x_j = 1 if an operation is located at node j, and 0 otherwise;
y_i = 1 if a demand node i is covered by at least one potential operational location j, and 0 otherwise;
$N_i = \{j \in J \mid d_{ij} \leq S\}$;
d_{ij} = the shortest distance between node i and node j;
S = maximum established value of the distance between the demand node and the operation node (desired service distance);
a_i = population served at demand node i;
P = number of operations to be allocated.

N_i is the set of operations sites eligible to provide coverage for a demand node i. A demand node is covered when the location of the operation closest to the node is less than or equal to S.

Analyzing the objective function described in Eq. (1) shows that the goal is to maximize the number of demand nodes covered within the desired service distance. The following equations present the constraints of the problem. Equation (2) lets y be equal to 1 only when one or more operations are established at locations within the set N (i.e., at one or more operations that are allocated within the distance S relative to the demand node i). Equation (3) defines the number of allocated facilities, which it is limited to by P, as defined by the DM. Finally, Eqs. (4) and (5) indicate that only integer values can be part of the solution, since the model is integer and uses binary programming.

The MCLP model requires some inputs to be made [8, 10, 13], e.g. the points of demand and possible local candidates to which a facility could be allocated. Demand points can be collected in the field, but candidate sites can be defined in different ways [5, 14], such as by using clustering methods [15, 16].

2.2 Clustering Methods and the K-Means Method

Data mining is a term used in the literature to describe the discovery of knowledge through databases. Therefore, data mining is a process that can make use of techniques such as artificial intelligence, statistics, mathematics and machine learning, to identify and extract useful information and subsequent knowledge from large databases [17].

According to [16], data classification is one of the tools used by data mining to identify and extract useful information. In the data mining process, the data is analyzed and classified by a common pattern or characteristic. Thus, clustering procedures are used to find these patterns, and aim to find a set of k points to act as centers and then assign each point to its nearest center, thus forming k clusters [15, 16, 18].

This procedure is essential in models aimed at the problem of locating n installations, since it can be synthesized when locating the k centroides, thus defining the possible candidate locations at which to position the installations [14, 16, 22]. According to [5], this pre-processing, to identify candidate locations, is essential for the execution of a model, and therefore this makes it possible to find a solution that meets most of the demand that the DM requires.

While several methods can be identified in the literature to perform this clustering, the k-means algorithm is the most used when it is desired to partition n measured quantities into k clusters [19–21]. Therefore, the k-means algorithm, can solve this pre-processing relatively quickly with regard to siting installations [16, 22].

The k-means algorithm was first introduced by [19] and is one of the unsupervised machine learning techniques that conducts the exploratory categorization of data until observations share a maximum similarity in the same cluster, with minimal similarity between different clusters [23].

According to [14] for a better understanding of the k-means method, some simplifications can be made to present the objective of the method in its mathematical structure. Thus, it can be presented with the cost function [14, 16, 24, 25] which contains all clusters and can be understood by Eq. (6).

$$\text{cost (kmeans)} = \sum_{k=1}^{k} \sum_{x_i \in C_k} d(x_i, x_{0k}) \tag{6}$$

Where x_{0k} is the cluster centroid C_k and $d(x_i, x_{0k})$ is the distance between point x_i and point x_{0k}. Therefore, the k-means algorithm seeks to minimize the sum of the distances between the given points (or points of demand, if a localization problem) to the nearest centroid [14, 16, 18, 19, 22, 25, 26]. Consequently, the k-means method can interactively structure a set of clusters from a data set [27].

[20] applied the k-means method in order to find candidate sites in an industrial plant in the food industry and found that the approach is consistent, especially for the models for dealing with the coverage problem. So, since k-means can be relatively scalable and efficient for large data sets, the method often ends up finding an optimal centroid, which minimizes the sum of the distances between the data points (the facilities and the demand points) [28].

Therefore, k-means can be added as a computational clustering tool in MIS which demands the handling of large volumes of data. It allows MIS to run more efficiently and, thus, helps DMs.

3 The MIS

The main objective of the MIS developed and proposed to solve the problem of this study is to provide a quick analysis so that the DM can increase his/her productivity when allocating police facilities. The maximum coverage of the facilities, in relation to the places where crimes occurred in the region, is taken as a guideline for the problem. The MIS was developed using the RStudio Software virtual environment.

The data stored in several databases are used to analyze several alternative locations. Therefore, these data can be transformed into information and knowledge by using the MIS.

Therefore, the proposed MIS architecture comprises the following elements:

- Software: is responsible for coordinating several groups of logic components that control the entire computer and the data processing system, thereby ensuring its operation. The main software programs are the Operating System, R and RStudio which help the various interactions of the DM with MIS.
- Hardware: encompasses all physical devices and equipment used for information processing, i.e. the central processing unit, the memory and the input and output devices of a given machine. The main hardware comprises: the computer, the modem to access the Internet and the various peripherals such as the printer, the monitor, the mouse and the keyboard.
- Data: are responsible for grouping and analyzing data related to public security from different geographical areas. The main data consist of the occurrences of crimes in the city of Recife that were obtained from the platform "Onde Fui Roubado" during the period from 2012 to 2017. Other data include the parameters required by the location model, such as the maximum coverage number of facilities to be allocated, which are defined by the DM.
- Method: this is responsible for grouping and processing all the methods for carrying out the work together, ensuring speed and confidence in processing the calculations when finding the best solutions to the location problem. Thus, the main methods are: a MCLP problem processing model, a K-means clustering model and a georeferencing model.
- People: the main people for the proposed MIS are DMs and experts. Where DMs are in charge of using the MIS in search of suggestions for allocating police facilities and the experts who are responsible for working on data processing and the mathematical models involved in MIS.
- Report with location suggestions for police facilities is responsible for compiling the reports that are generated by the MIS and allowing the DM to download them. Where the main reports are directed to candidate locations, selected locations, final allocation result and sensitivity analysis.

The proposed MIS is divided into 8 intuitive, user interaction screens, where each screen will be called a page. Thus, there are the Files, Data, K-means, Candidate Locations, Allocation (MCLP), Download, About, and Settings pages. In the following figures, some pages will be presented (MIS functionalities). In Fig. 1 the File page is illustrated. It is responsible for facilitating user access when importing data into MIS. Thus, it is possible for the user to format and display how the data will be imported into the system.

At one-point MIS requires the number of clusters that the DM wants to partition the database of crimes. Crimes are grouped into clusters using the k-means model, similarly to the region where crimes most often occur, as shown in Fig. 2. Thus, by using the k-means tool, the centroid of each of the clusters can be found and worked with as a candidate location to allocate a police facility. If the numbers of optimal locations are

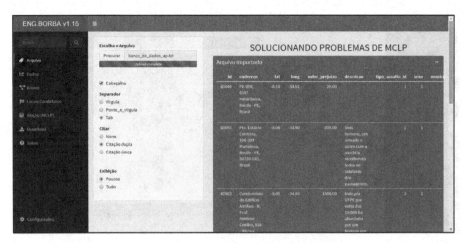

Fig. 1. File page of the MIS proposed.

not adequate due to the number of clusters available, some adjustments can be made to the MIS so that it complies with the number of clusters that the DM needs.

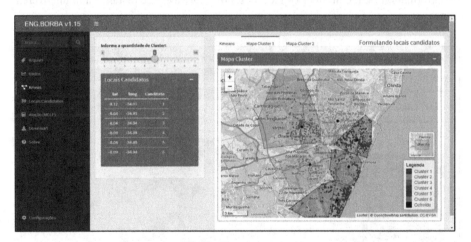

Fig. 2. K-means page.

The Candidate Places page is illustrated in Fig. 3, which offers an analysis of the crimes that will be covered by an operation area. Thus, the page has a set of four simultaneous analyses ("Max Coverage", "Min Not Covered", "Average Distance" and "Average Deviation"), which allows the user to gain a broad understanding of the problem being faced. This provides better direction in decision making to establish the number of facilities to be allocated and the area to be covered.

In Fig. 4 the Allocation page (MCLP) is presented and it shows the coordinates that were selected to install a police unit among the coordinates of the candidate locations. For this purpose, information provided by the user regarding the range of coverage and

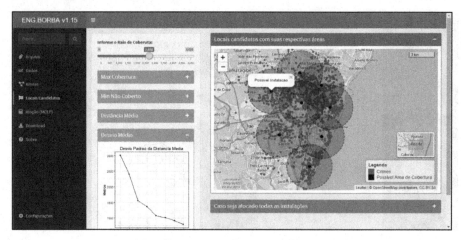

Fig. 3. Candidate Places page.

the number of facilities to be allocated was used to maximize the volume of crimes that will be covered. This gives a quick and clear result of the facility allocation, with numerical information on the number of facilities, coverage, average distance and uncovered crimes.

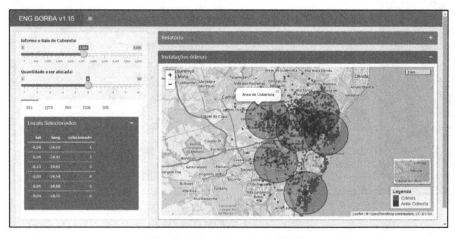

Fig. 4. Allocation page (MCLP).

The MIS also provides some reports of the analyses performed. And finally, it becomes apparent that the user finds it easy to interact with the proposed MIS to solve the problem of this study. The analysis boxes, selection boxes, maps, and buttons integrated into the MIS interface enables anyone to handle it. And so, it becomes easier to develop the case study of this research, which is described below.

4   The Case Study

The case study carried out was applied in a real case where the problem of siting police installations is fundamental in order to reduce the crime rates and improve meeting the existing demand. The place where the research was applied was the city of Recife, capital of the state of Pernambuco, which is located in the Northeast of Brazil.

The database used in this research was obtained from the platform "Where I was robbed", which is a collaborative platform against violence. On the platform, anonymous users mark on a map the places in which they suffered theft, robbery, lightning kidnapping, robbery from their homes, from shops and other types of crime.

4.1   Candidate Locations

Before accurately and optimally locating police facilities in the territory of the City of Recife, an initial study needs to be done to determine which candidate locations will be allocated a police installation. Thus, the proposed MIS and the database under study were used to group crimes into clusters by proximity to the region where crimes are occurring most frequently. Thus, Fig. 5 illustrates the ten regions where crimes are occurring most frequently and similarly in the city of Recife.

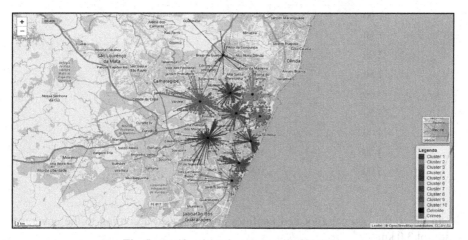

Fig. 5. Recife crime clusters provided by MIS.

As a specific and exact coverage radius is not provided for the facilities of the city of Recife, this study presents several scenarios ranging from one to five kilometers of coverage. To analyze the sensitivity of this parameter, the behavior of the optimized allocation of these facilities, for each coverage radius, was verified.

4.2 Optimal Allocation

After defining the candidate locations and the range of values regarding the coverage radius, the budgetary constraint, i.e. the number of facilities that will be allocated, must be defined.

Thus, out of the ten candidate sites, no more than five facilities were selected and tested. This choice of only five installations was due to the initial interest in analyzing the optimal scenario obtained with the proposed MIS and the five Recife police facilities currently in operation.

At this stage it is important to highlight that to solve the problem presented, which consists of 1688 points of demand and 10 candidate sites, the MIS required approximately 3.09 s.

Thus, we obtained the data from Table 1, which presents the gain in coverage, by increasing the operating radius of the five allocated facilities. It is clear that the radius values of 2.5, 3.0, and 3.5 km provide good coverage of the occurrences, i.e. 81.8%, 90.1% and 96.6%, respectively.

Table 1. Optimal coverage for the five allocated facilities.

Coverage radius (km)	Number of covered	Percent covered
1.00	582	34.5%
1.50	923	54.7%
2.00	1143	67.7%
2.50	1380	81.8%
3.00	1521	90.1%
3.50	1630	96.6%
4.00	1664	98.6%
4.50	1675	99.2%
5.00	1683	99.7%

In Fig. 6 the optimal allocated facilities are illustrated according to the coverage radius of 3.0 km.

4.3 Comparison of Performance

This section will present a comparison between performance of the current scenario (with current police unit locations) and possible performance with unit locations suggested by MIS. To this end, a scenario of five installations with a radius of three kilometers of coverage was defined.

In Fig. 7, the gain of coverage of occurrences by installations is presented, i.e. by adding new police units in the city of Recife, more crimes are covered by these facilities. The extra gain in coverage that the proposed model can achieve in relation to

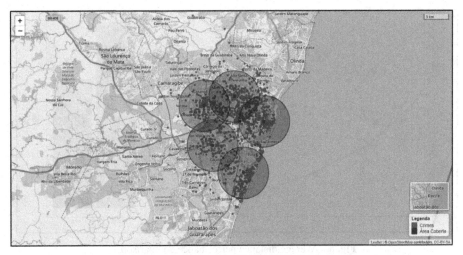

Fig. 6. Optimal allocated facilities within 3.0 km radius.

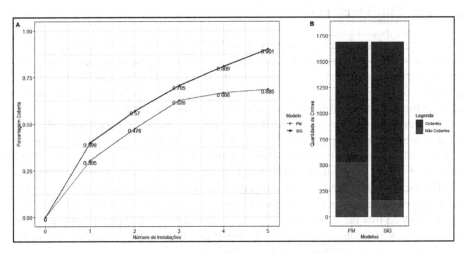

Fig. 7. Coverage performance comparison.

the current configuration is noteworthy. Using the suggested locations, a 90% coverage of occurrences can be achieved after the fifth facility has been allocated which represents a 22% improvement over the current scenario.

Figure 8 shows the average distance in meters from occurrences to the nearest facility, so that for each installation that is added to the territory of Recife, a significant reduction in this distance is obtained.

Using the units suggested by MIS, it is possible to achieve a reduction difference from the fifth allocated facility of almost 920 m away, considering the current configuration of 5 units.

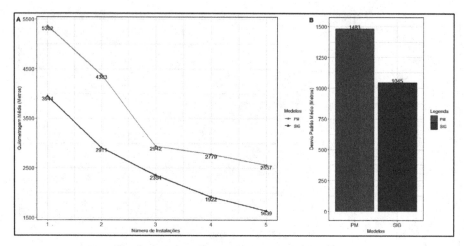

Fig. 8. Average distance performance comparison.

5 Final Considerations

The objective of this paper was to present an MIS developed for police unit locations, thereby aiming to increase operational efficiency in responding to the occurrence of a crime. The K-means model as a starting point in the search to find the candidate locations proved to be a great complement to the MCLP model, and together they were used to develop the GIS proposed in this research using RStudio software.

After developing the proposed MIS and its application in the city of Recife-PE, with the data available on the platform "Where I was robbed", it can be said that the general objective of this research was successfully achieved. Overall, the research has developed an innovative tool for modeling and solving the problem of locating police facilities.

The locations proposed by MIS suggest a better performance when compared to the performance of the current configuration of the police facilities of the city of Recife. The results showed that the proposed MIS exceeds the current configuration of the police units, in terms of location, by 22% of covered area and provides a reduction of almost 920 m in the average distance to cover the incident.

It is noteworthy that the proposed MIS can be adjusted and include new features to better meet a DM's needs. It is also suggested this research be continued by using a new programming language, which facilitates the insertion of new modules and variables.

Acknowledgment. The authors would like to acknowledge the support received to develop this research from the National Council for the Improvement of Higher Education (CAPES), a foundation affiliated with the Ministry of Education in Brazil, and the Brazilian National Research Council (CNPq) (Process number 307344/2017-7 and Process number: 422785/2016-4 - CHAMADA UNIVERSAL MCTI/CNPq N° 01/2016).

References

1. Sohngen, C., Cipriani, M.: Crime E Violência No Brasil: Representações Sociocultur-ais Na Pós-Modernidade [Crime and violence in Brazil: sociocultural representations in postmodernity]. R. Inter. Interdisc. **16**(1), 36–56 (2019)
2. Anuário Brasileiro de Segurança Pública [Brazilian Yearbook of Public Security], 12° Edição. Fórum Brasileiro de Segurança Pública, São Paulo (2018). ISSN 1983-7364. http://www.forumseguranca.org.br/publicacoes/anuario-brasileiro-de-seguranca-publica-2018
3. Mitchell, P.: Optimal selection of police patrol beats. J. Crim. Law Criminol. Police Sci. **63**(4), 577–584 (1972)
4. Henriques de Gusmão, A.P., Aragão Pereira, R.M., Silva, M.M., da Costa Borba, B.F.: The use of a decision support system to aid a location problem regarding a public security facility. In: Freitas, P.S.A., Dargam, F., Moreno, J.M. (eds.) EmC-ICDSST 2019. LNBIP, vol. 348, pp. 15–27. Springer, Cham (2019). https://doi.org/10.1007/978-3-030-18819-1_2
5. Revelle, C.S., Eiselt, H.A.: Location analysis: a synthesis and survey. Eur. J. Oper. Res. **165**, 1–19 (2005)
6. Morais Gurgel, A., Pires Ferreira, R.J., Aloise, D.J.: Proposta de modelos para a localização de bases policiais e roteirização de viaturas [Proposed models for the location of police bases and vehicle routing]. ENEGEP (2010)
7. Pizzolato, N.D., Raupp, F.M.P., Alzamora, G.S.: Revisão de desafios aplicados em local-ização com base em modelos da p-mediana e suas variantes [Review of challenges applied in localization based on p-median models and their variants]. Pesquisa Operacional para o Desenvolvimento [Oper. Res. Dev.] **4**(1), 13–42 (2012)
8. Owen, S.H., Daskin, M.S.: Strategic facility location: a review. Eur. J. Oper. Res. **111**, 423–447 (1998)
9. Fallah, H., Naimisadigh, A., Aslanzadeh, M.: Covering problem. In: Zanjirani, Farahani R., Hekmatfar, M. (eds.) Facility Location: Concepts, Models, Algorithms and Case Studies. Contributions to Management Science. Physica Verlag, Heidelberg (2009). https://doi.org/10.1007/978-3-7908-2151-2_7
10. Church, R.L., Revelle, C.: The maximal covering location problem. Pap. Reg. Sci. Assoc. **32**, 101–118 (1974)
11. Cordeau, J.-F., Furini, F., Ljubic, I.: Benders decomposition for very large-scale partial set covering and maximal covering location problems. Eur. J. Oper. Res. **275**, 882–896 (2018)
12. Murray, A.T.: Maximal coverage location problem: impacts, significance, and evolution. Int. Reg. Sci. Rev. **39**(1), 5–27 (2016)
13. Hakimi, S.L.: Optimum locations of switching centers and the absolute centers and medians of a graph. Oper. Res. **12**, 450–459 (1964)
14. Silva, V.F., Cabral, L.A.F., Quirno, R.: Estratégia para seleção de locais candidatos voltada ao problema de localização de viaturas policiais [Strategy for the selection of candidate sites focused on the problem of locating police vehicles]. In: XLV SBPO- 19 Simpósio Brasileiro de Pesquisa Operacional, Natal/RN, setembro de 2013
15. Kaveh, P., Zadeh, S.A., Sahraeian, R.: Solving capacitated P-median problem by hybrid K-means clustering and FNS algorithm. Int. J. Innov. Manag. Technol. **1**(4), 405–410 (2010). Industrial Engineering and Operations Management, Dhaka, Bangladesh
16. Oliveira, M.G.: Sistema de Localização de Facilidades: Uma abordagem para mensuração de pontos de demanda e localização de facilidades [Facilities location system: an approach for measuring demand points and facility location]. Dissertação apresentada ao Programa de Pós-Graduação do Instituto de Informática da Universidade Federal de Goiás, Goiânia (2012)
17. Turban, E., Aronson, J.E., Liang, T.-P.: Decision Support System and Intelligent System, 7th edn. Prentice Hall, Upper Saddle River (2005)

18. David, S., Shwartz, S.S.: Understanding Machine Learning: From Theory to Algorithms, vol. 1, pp. 1–449. Cambridge University Press, Cambridge (2014)
19. Macqueen, J.: Some methods for classification and analysis of multivariate observations. In: Proceedings of the Fifth Berkeley Symposium on Mathematical Statistics and Probability, vol. 1, no. 14, pp. 281–297 (1967)
20. Hartigan, J.A.: Clustering Algorithms, pp. 113–129. Wiley, New York (1975)
21. Praveen Kumar, D., Amgoth, T., Annavarapu, C.S.R.: Machine learning algorithms for wireless sensor networks: a survey. Inform. Fusion **49**, 1–25 (2019)
22. Risso, L.A., et al.: Clusterização K-Means Para Decisão De Localização Industrial [K-means clustering for industrial location decision]. In: XXXV ENCONTRO NACIONAL DE ENGENHARIA DE PRODUCAO, Fortaleza, CE (2015)
23. Koo, C., et al.: A novel estimation approach for the solar radiation potential with its complex spatial pattern via machine-learning techniques. Renewable Energy **133**, 575–592 (2019)
24. Hair, J.F., Rolph, A.E., Tatham, R.L., Black, W.C.: Análise Multivariada de Dados [Multivariate Data Analysis], 5th edn. Bookman, Porto Alegre (2005)
25. Friggstad, Z., Khodamoradi, K., Rezapour, M., Salavatipour, M.R.: Approximation schemes for clustering with outliers. ACM Trans. Algorithms (TALG) **15**(2), 1–26 (2019)
26. Lantz, B.: Machine Learning with R. Discover How to Build Machine Learning Algorithms, Prepare Data, and Dig Deep into Data Prediction Techniques with R, 2nd edn, pp. 1–452. Packt Publishing, Birmingham (2015)
27. Kanungo, T., et al.: An efficient k-means clustering algorithm: analysis and implementation. IEEE Trans. Pattern Anal. Mach. Intell. **24**, 881–892 (2002)
28. Berkhin, P.: Survey of Clustering Data Mining Techniques. Accrue Software, Inc. (2002)

An Automated Corpus Annotation Experiment in Brazilian Portuguese for Sentiment Analysis in Public Security

Victor Diogho Heuer de Carvalho[1,2(✉)], Thyago Celso Cavalcante Nepomuceno[3], and Ana Paula Cabral Seixas Costa[2]

[1] Universidade Federal de Alagoas, Delmiro Gouveia, AL, Brazil
[2] Universidade Federal de Pernambuco, Recife, PE, Brazil
`victor.hcarvalho@ufpe.br, apcabral@cdsid.org.br`
[3] Universidade Federal de Pernambuco, Caruaru, PE, Brazil
`thyago.nepomuceno@ufpe.br`

Abstract. This paper aims to present an experiment developed in order to produce a corpus with automated annotation, using pre-existing annotated corpus and machine learning classification methods. A search for pre-existing annotated corpora in Brazilian Portuguese was applied, founding six corpora of which one has been selected as the training dataset. A set of tweets was collected in a specific area of Recife (Pernambuco-Brazil) using some keywords related to kinds of crimes and reinforcing some places in that area. Preprocessing tasks were applied over the pre-existing corpus and the tweets' set collected. Latent Dirichlet Allocation was applied for topic modeling followed by Multinomial Naïve Bayes, Linear Support Vector Machines, and Logistic Regression for the sentiment polarity classification. The results of the cross-validation of the experiment indicated Linear Support Vector Machines as the most accurate classification method among the three considering the specific training set used, and by this method, the new annotated corpus about the selected topic related to public security was created.

Keywords: Corpus annotation · Sentiment analysis · Public security · Brazilian Portuguese · Machine learning classification

1 Introduction

The social web has added a new level of challenge for organizations that want to collect data from this massive source to apply in their decision-making [1]. Big data and social media analytics have become necessary concepts for companies to remain competitive while meeting the expectations of their customers/consumers [2]. From these concepts arises the need to analyze the massive and heterogeneous volumes of data from the social web [3] in order to provide more accurate information based on public opinion decision-makers [4].

Daily social network users produce textual records that can be useful for organizations if properly handled, and text mining is essential to ensure these records can generate

© Springer Nature Switzerland AG 2020
J. M. Moreno-Jiménez et al. (Eds.): ICDSST 2020, LNBIP 384, pp. 99–111, 2020.
https://doi.org/10.1007/978-3-030-46224-6_8

useful information [5, 6]. Web mining is a text mining approach that aims to explore and extract data from various sources across the internet, enabling pattern recognition and the extraction of useful information [7]. As an intermediate outcome, sentiment analysis, which in turn is a specific type of natural language processing problem, can be used to classify the polarity of textually expressed opinions [4, 8].

Public services management agencies are potential users of the tools and approaches mentioned above, highlighting applications in health, environment, and security [9]. Public security agencies, for instance, can use social web mining to extract people's impressions about security policies and actions from social networks aggregating these information to reports on crimes and policing in specific periods and geographical regions, assisting decisions on measures to improve these policies and actions [10] or crossing the results of the analytical process with internal data to support decisions on improving surveillance by predicting events against security [11].

Assessing people's sentiments about public security is a process of interest to government public security management agencies. General and specific misunderstandings have been recurring topics on the authorities' agenda, and many recent studies are devoted to discussing the empirical context of crime in the environment in which such data collection is done [12–14].

Sentiment analysis requires a series of preparations so that the opinions expressed by people can be classified appropriately and reliably to the sentiment they really wanted to express in their textual records [15]. The creation and use of an adequately annotated corpus are fundamental for the classification of sentiments polarities as the main result of the analytical process [16].

The annotation process can be done manually, using individuals with enough expertise to analyze the texts and apply a polarity label [17, 18], or using automated procedures based on machine learning techniques which depends on a pre-existing annotated corpus and computational capacity to process the new corpus in a real-time [19, 20].

Topic modeling is another important task to ensure obtaining a specific domain corpus. It allows the most recurring words in the text to be identified and clustered so that each cluster defines a topic that can also be annotated in a corpus, allowing the extraction of specific textual records for a topic of interest [21].

This paper aims to present a corpus annotation experiment dedicated specifically to a process of sentiment analysis about Public Security in the city of Recife (Brazil). To this end, a set of tweets was collected in a specific area of the city, using some key terms related to crimes and reinforcing some places in that area. Topic modeling was performed to identify the tweets most adjusted to the public security theme and, afterward, an automated procedure was applied using a pre-existing corpus for the annotation of the new (specialized) corpus. Obtaining a specialized annotated corpus is a necessary process to work on a specific domain, as occurs on the broader context in which this paper is inserted.

The sequence is divided as follows: Sect. 2 provides a background on the corpus annotation process; Sect. 3 presents the annotation experiment process applied; Sect. 4 presents the experiment results; lastly, Sect. 5 presents the concluding remarks containing some practical implications and indications of future work.

2 Background

The following subsections introduce some main concepts about the automated corpus annotation process, topic modeling, and sentiments polarity labeling.

2.1 Automated Corpus Annotation

Automated corpus annotation presupposes the use of a pre-existing annotated and general-purpose corpus as a training dataset for machine learning algorithms to perform polarity labeling over a new domain-specific corpus [22].

It is important to emphasize that neither manual and automated processes are infallible and both have advantages and disadvantages [23]: (i) manual annotation can be executed with little corpus preparation, and ensure the best accuracy in the results since it is a process-oriented to human perceptions; in contrast, the process is slow and limited to few results; (ii) automated annotation can work with broader corpus and the process is much faster compared to manual, but it involves the programming of a labeling or tagging process using artificial intelligence techniques that should be first trained, so, the accuracy of the results will depend on the quality of the pre-existing training set.

For automated annotation, sometimes, several pre-existing corpora can be used to ensure more accurate classifications of the sentiments [20]. Besides, some review and re-annotation processes may be necessary to ensure the quality of the new corpus [19]. Preprocessing techniques also are required for the preparation of the corpus as well as the application of cross-validation using metrics like accuracy, precision, recall, and F1-score [18, 20, 24, 25].

2.2 Topic Modeling

The topic modeling allows the classification/labeling of texts according to the topics found [26], ensuring the identification of discussion subjects in the texts [27], supporting the creation of a new corpus.

Latent Dirichlet Allocation (LDA) figures as a recurring topic modeling technique in recent literature (see, for instance, [27–30]). It belongs to the class of hierarchical Bayesian models describing documents as a mixture of topics [29] and still can be classified as an unsupervised machine learning generative technique [31]. LDA was developed to fix issues related to Latent Semantic Analysis (LSA) and its probabilistic version (PLSA) [32], two other topic modeling techniques based on the use of a general matrix of texts and terms to be decomposed in other two matrixes with the relations "document to topic" and "topic to term" [31].

2.3 Sentiment Polarity Labeling

Sentiment polarity labeling is a process that can be performed by machine learning techniques, lexicon-based methods, or hybrid approaches [33]. The machine learning techniques applied to sentiment analysis are designated to train text classifiers, according to pre-existing datasets with polarity annotation [25].

The most recurrent machine learning techniques for this purpose on literature are the classic trio Naïve Bayes, Support Vector Machines (SVM) and Maximum Entropy/Logistic regression followed by artificial neural networks and deep learning [34], but other supervised methods such as Random Forests, Nearest Neighbors, as well as unsupervised methods as K-Means [33], for instance, can be applied.

Lexicon based methods involve the calculation of the text semantic orientation to determine the polarity [16]. The approach involved on this process is oriented to counting and weighting the sentiment-related words, and it can be performed by three ways [25]: (i) a manual process, as mentioned before, depends on human beings and is time-consuming; (ii) a dictionary-based process explores lexicographical resources, as Word-Net, for instance; (iii) a corpus-based process uses sets of words with well-defined sentiment polarity exploring syntactic relations to identify new sentiment words.

Lastly, hybrid approaches may combine both machine learning/statistics methods and lexicon-based approaches [16], helping to improve the accuracy of sentiments classification [35, 36]. The next section will present the steps of the experiment applied to perform corpus annotation using a pre-existing annotated corpus in Brazilian Portuguese, and tweets collected from a specific area in a large Brazilian city.

3   Experiment Description

The corpus annotation experiment adopted in this work was based on automated tasks. It involved as the first task the search for a pre-existing annotated general-purpose corpus in Brazilian Portuguese and the collection of a set of tweets in a specific area from a Brazilian city, using some initial filtering based on keywords and setting the collector only to get tweets in the target language.

The next task was to apply topic modeling to identify which topics were, in fact, relevant for obtaining the corpus dedicated to public security sentiment analysis. Lastly, using the pre-existing corpus and the tweets extracted based on topic modeling, sentiment analysis was applied to obtain the final specific annotated corpus in Brazilian Portuguese. Figure 1 presents the workflow of this process.

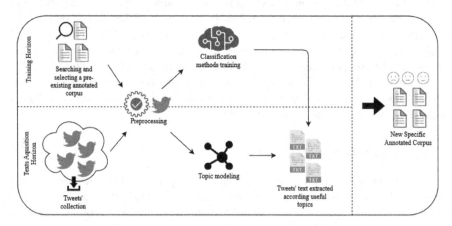

Fig. 1. Annotation experiment process workflow.

Each element in this workflow will be described in the following subsections, demonstrating what technologies and methods were used.

3.1 Searching and Selecting a Pre-existing Annotated Corpus

Here, pre-existing annotated corpora obtained through well-structured processes were sought to select a textual training data set for machine learning algorithms applied in the classification of the sentiments' polarities. This search found the corpora related to the following works: [17, 37–41]. One of the corpora [38] was discarded for not presenting sentiments polarity labeling. Another corpus [39] was based on a bookshelf only applied positive and negative labels, being eliminated as it was desired to consider a neutral label too. Among the remaining corpus, the one related to the UniLex method [40] was chosen, since it fits well with the classification proposal.

The selected corpus consisted of fifteen XLSX files that were read through a Python script using the Glob library to read multiple files and Pandas to load the data from these files into a data frame and convert this structure to a final file in CSV format. The texts on the final file were preprocessed by a function that uses some methods from Natural Language Toolkit (NLTK) [42] and "re" (regular expressions) libraries for data cleaning and stop words removal, using Portuguese stop words downloaded via NLTK. The preprocessing tasks ensure noises elimination in the training of the classification methods. More details about the selected corpus are presented in the results section.

3.2 Tweets' Collection and Preprocessing

Tweet's collection was based on the procedure to collect users' posts on Twitter's continuous stream, storing them in a JSON file [43]. A stream collector script was implemented using Python language applying the following libraries and their methods: Tweepy for accessing the Twitter API, "time" providing function for dealing with time elements, and "os" for miscellaneous interface with the operating system. The JSON content was charged in Pandas' data frames to be manipulated by other scripts.

Preparing the tweets' data for the topic modeling and polarity labeling tasks is an element of significant interest in the whole analytical process since its focus is to give a suitable structure so that the texts can be effectively analyzed. For this purpose, duplicate tweets were dropped, and the same preprocessing function applied to the selected corpus was used to preprocess the tweets' texts also to eliminate noise for the subsequent parts of the analysis.

3.3 Topic Modeling

LDA was the method selected for topic modeling, following the tendency presented in the background section. Reading the works in the previous section related to LDA applications is recommended to ensure understanding of how the method works. A new Python script was implemented to apply topic modeling, this time using Scikit-Learn library calling functions for feature selection (TF-IDF, term frequency-inverse document frequency), LDA functions, and Grid Search functions to select the best topic model according to LDA results and the feature selection applied.

Another scientific programming and graph plotting supporting libraries like Numpy, Matplotlib, and Seaborn also were used. Plotting libraries allowed the visualization of the most recurrent textual features selected (bigrams), and of the topics segmentation as texts' clusters. The k-Means technique, from Scikit-Learn, was used here for the sole purpose of segmenting these topics into the related texts according to the labeling done by the LDA, generating the graph with the corresponding visualization.

With topic modeling, the tweets could be labeled based on the best topic model found, and the final selection of tweets searched for texts related to public security issues occurred, defining the subset of texts for polarity labeling.

3.4 Polarity Labeling Over the Selected Tweets

The polarity labeling is the final part of the procedure, dedicated to classifying the sentiments' polarities using some supervised machine learning classification techniques. The steps presented in Subsects. 3.1, 3.2, and 3.3 provided (i) a pre-existing annotated corpus, (ii) a broader set of unclassified tweets, and (iii) a subset of unclassified tweets extracted according to a public security-related topic identified using LDA.

The pre-existing annotated corpus contains the sentiments' polarities manually annotated using three labels: negative (-1), neutral (0), and positive (1). It was used to train the machine learning techniques, applying cross-validation, and allowing to evaluate the sentiment classification according to specific metrics [25] supporting discovering which is the most suitable technique for this task among those used. It is important to emphasize that only the tweets in the subset extracted using LDA were used for applying the automated polarity labeling/annotation, using the classification techniques.

The last Python script was implemented to process polarity labeling, using the three classification techniques from Scikit-Learn: Multinomial Naïve Bayes, Linear Support Vector Machines (Linear SVM), and Logistic Regression. The results of the described process will be presented following.

4   Experiment Results

4.1 Pre-existing Annotated Corpus Characteristics

The pre-existing corpus, derived from the UniLex method [40], is composed of 12668 tweets extracted in Brazilian Portuguese, related to issues about national politics. It is openly available to be used in sentiments analysis processes. The corpus structure is composed of 3 columns, the first containing the index o the tweets, the second the tweets, and the third the labels classifying the tweets (negative, neutral, and positive). The texts are distributed in 4197 negatives, 4753 neutrals, and 3715 positive registers.

No duplicate entries were found, however, three invalid entries were found and needed to be deleted, resulting in an amount of 12665 tweets with the same amounts for each polarity label. Table 1 exemplifies the first three registers of the corpus after preprocessing with tweets' texts in Brazilian Portuguese (index column was added in the initial corpus to data frame conversion).

Data preprocessing eliminated special characters, the "RT" (retweet) indication, entity tags with "@", emojis and stop words from Brazilian Portuguese, reducing noises with unnecessary textual elements during the training of the classification methods.

Table 1. Example of the corpus structure.

Index	Tweets	Labels
0	#caonossodecadadia #novo vanessa mandotti dir pgm começo	Neutral (0)
1	bola frente amanhã outro dia, outra cena, outra chance, viver ser feliz irmão amém #amanha #novo #viver	Positive (1)
2	cara mal? acho apenas corte diferente barba #visu #novo #gostou	Positive (1)

4.2 Extracted Tweets Characteristics

The stream tweets collector script was parameterized to perform the searches in a specific quadrant of the city of Recife, on the following coordinates: -34.9090 for west latitude, -8.0716 for south longitude, -34.8753 for east latitude and -8.0522 for north longitude. In addition to these geographical delimiters, the specific language of the collection was defined to Portuguese, and some keywords were introduced for filtering the initial collection: *["tráfico de drogas", "estupro", "homicídio", "assassinato", "furto", "roubo", "assalto", "Soledade", "Derby", "Boa Vista", "Madalena", "Ilha do Retiro", "Paissandu", "Ilha do Leite", "Joana Bezerra", "Santo Antônio", "Santo Amaro", "Coque", "Borel", "Ilha de Deus", "Antigo", "Pina"]*. These keywords are related to some common crimes' designations in Portuguese and some places, to reinforce the tweets' searches related to the defined Recife's quadrant.

The number of tweets collected was 17218, but after duplicities exclusion, the number was reduced to 10283. These tweets were stored in a JSON file, and again, the same preprocessing tasks used before were applied.

4.3 Topic Modeling Results

After preprocessing the tweets collected on the defined Recife's quadrant, the topic modeling process was performed. TF-IDF using bigrams was applied instead of simple counting, to create the bag-of-words and discover the frequency of the bigrams among the tweets' texts. Figure 2 contains a bar plot with the ten most common bigrams found on the text set.

A Grid Search was applied to find the best LDA model according to the number of topics in an analysis using 5, 10, 15, 20, 25, and 30 topics, and 0.5, 0.7, and 0.9 learning decays. The results indicated the best log-likelihood score (-144000.0555) and model perplexity (408562.2158) for a model with five topics and a learning decay of 0.7. As higher the log-likelihood and as lower the model perplexity, the best the model [44]. The model was formulated to contain the ten most relevant bigrams extracted from the tweet texts for each topic.

The topics in the texts' set follow the distribution presented in Table 2, where it can be noted that topic 0 is the most frequent. The occurrence numbers in this table were determined using the function *values_count* from Pandas (Python's library), according to the topics extracted through LDA.

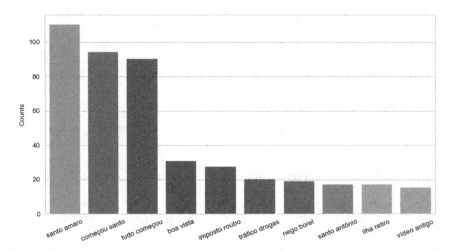

Fig. 2. The ten most common bigrams on tweets' texts.

Table 2. The occurrence numbers of each topic on the model.

Topic	Occurrences
0	2278
3	2151
4	1997
1	1958
2	1899
Total	10283

Figure 3 presents the segmentation of the topics (clusters) on the tweets' set, according to the components' weights, using the K-Means technique.

Finally, only tweets related to "Topic 0" were extracted in a new data frame and converted to a CSV file, to proceed with the last part of the experiment. This topic seems to be related to some crimes (robbery, murder, and drug trafficking) that took place in the vicinity of the central area of Recife. The top 10 bigrams in this topic are: *["mulheres manifestações", "vou fazer", "violência contra", "tentativa homicídio", "ilha retiro", "gente caso", "boa vista", "tráfico drogas", "nego borel", "amor antigo"].*

4.4 Polarity Labeling and Public Security Related Corpus Obtention

The last part of the experiment started with using the preprocessing results from the pre-existing corpus with sentiment labeling to train the three classification methods: Multinomial Naïve Bayes, Linear SVM, and Logistic Regression. Again, a bag-of-words was created using TF-IDF, but this time with unigrams.

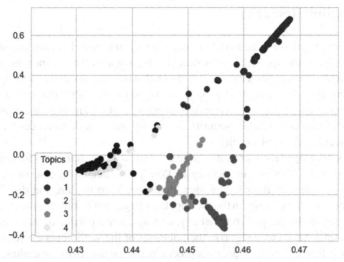

Fig. 3. Topics (clusters) segmentation.

Table 3 presented the summary of the cross-validation of the methods, using accuracy for each method in general and precision, recall, and F1-score metrics for each polarity in each method.

Table 3. Methods metrics summary related to the training corpus.

Methods	Multinomial Naïve Bayes			Linear SVM			Logistic regression		
Metric	Polarity								
	−1	0	1	−1	0	1	−1	0	1
Accuracy	0.5507			0.5533			0.5492		
Precision	0.53	0.60	0.51	0.58	0.57	0.48	0.58	0.56	0.48
Recall	0.77	0.56	0.28	0.64	0.60	0.40	0.63	0.61	0.37
F1-score	0.63	0.58	0.36	0.61	0.58	0.44	0.60	0.59	0.42

This table supported the final method selection for the corpus related to a Public Security topic annotation. Linear SVM presented the best accuracy on the cross-validation, with 0.5533 against 0.5507 from Multinomial Naïve Bayes (the second-best) and 0.5492 from Logistic Regression. Thus, for the experiment, the Linear SVM was selected as the final training method for the new corpus.

The new corpus was saved in a CSV file, and tests applying the cross-validation over it as a new training set presented an accuracy of 0.7405 for Logistic Regression, 0.6527 for Multinomial Naïve Bayes and 0.7379 for Linear SVM.

5   Concluding Remarks

This paper presented some essential concepts about automated corpus annotation for sentiment analysis and applied an experiment using some of these concepts. Sentiment analysis represents a central element in a research that is being developed with application in public security in the city of Recife (Brazil). As the main result, a corpus was obtained from the applied experiment for use as a training dataset to automate the classification of tweets' polarities about public security events, situations or occurrences with Twitter users in a specific region of the city.

The work developed has its novelty in the creation of a specific corpus for applications of sentiment analysis in public security in Portuguese language, also dedicated to a specific urban region in Brazil. The idea, therefore, is to explore the distribution of sentiment about public security in the Metropolitan Region of Recife from textual records of tweets published by people from that city, writing down a portion of these records and using it as a training set to categorize the polarities of the sentiments expressed in new texts also from this region. This does not preclude that the described procedure is used in other geographic regions and other fields such as public education, transportation, and health.

Future studies should be developed using computational power higher than that used in the reported experiment, and depending on the amount of data, prepared to deal with Big Data strategies. Other methods can be tested in both topic modeling, sentiment polarity classification, and further searches about pre-existing annotated corpora in the Portuguese language will be done. A more refined assessment of sentiment polarity classification methods will also be made, based on all metrics used, to assist in the ranking and selection of the best method.

For those interested in applying the process presented in the experiment in other languages, it is recommended to change the search language in the searching tool. Changes in the geographic region where the collection of tweets will be applied will also imply language changes, as well as the determination of key-terms for the search in the target language. For instance, in the experiment reported in this work, these parameters were determined in a Twitter stream collector script written in Python and can be easily changed to do searches in other languages or regions.

For training sets in other languages, those interested can search for corpora in the target language. Another option is the use of some automated translation process, taking NLTK, which contains translation functions (nltk.translation), as an example.

Other treatments regarding irony/sarcasm and vagueness should also be applied to improve the process of obtaining new corpora for use in research. It is also intended to use geo-referenced information from social network records, when available, for the classification of urban areas regarding the sentiments of their inhabitants about public security events and measures. Instead of using only metrics like precision, accuracy, recall, and F1-score, the Receiver Operating Characteristic (ROC) curve can be applied too, allowing the visualization of the general performances of each method, assisting in the selection of the most adjusted.

The results of the analyses carried out in this process are addressed to support activities from the Secretariat of Social Defense of Pernambuco since the research being developed is part of an agreement between the state government and the Federal University of Pernambuco, where the involved researchers are located. The process will also be incorporated into a sentiment visualization dashboard, allowing governmental managers to visualize public sentiment distribution over time and geographic space. This dashboard was planned as a module of a decision support system that is also being developed as an outcome for the referred agreement.

The sentiment visualization dashboard should help managers to combine time and geographic location to analyze the evolution of people's opinions registered in social networks, extracted through sentiment mining and analysis, and combining it, for instance, with other internal and external information about policing and crime occurrences to discover areas that demand more considerable attention for security actions.

Acknowledgment. This paper was funded in part by the Coordination for the Improvement of Higher Education Personnel (Brazil) – Finance Code 001, and by the National Council for Scientific and Technological Development (Brazil).

References

1. He, W., Wang, F.K., Akula, V.: Managing extracted knowledge from big social media data for business decision making. J. Knowl. Manage **21**, 275–294 (2017). https://doi.org/10.1108/JKM-07-2015-0296
2. Vatrapu, R., Mukkamala, R.R., Hussain, A., Flesch, B.: Social set analysis: a set theoretical approach to big data analytics. IEEE Access **4**, 2542–2571 (2016). https://doi.org/10.1109/ACCESS.2016.2559584
3. Colombo, P., Ferrari, E.: Access control in the era of big data: state of the art and research directions. In: Proceedings of the 23rd ACM on Symposium on Access Control Models and Technologies – SACMAT 2018, pp 185–192. ACM Press, New York, NY, USA (2018)
4. Bjurstrom, S.: Sentiment analysis methodology for social web intelligence. In: Proceedings of the Twenty-first Americas Conference on Information Systems. Association for Information Systems, Puerto Rico, pp 1–12 (2015)
5. Stieglitz, S., Mirbabaie, M., Ross, B., Neuberger, C.: Social media analytics – challenges in topic discovery, data collection, and data preparation. Int. J. Inf. Manage. **39**, 156–168 (2018). https://doi.org/10.1016/j.ijinfomgt.2017.12.002
6. Feng, L., Chiam, Y.K., Lo, S.K.: Text-mining techniques and tools for systematic literature reviews: a systematic literature review. In: 2017 24th Asia-Pacific Software Engineering Conference (APSEC), pp 41–50. IEEE (2017)
7. Lorentzen, D.G.: Webometrics benefitting from web mining? An investigation of methods and applications of two research fields. Scientometrics **99**, 409–445 (2014). https://doi.org/10.1007/s11192-013-1227-x
8. Sisodia, D.S., Reddy, N.R.: Sentiment analysis of prospective buyers of mega online sale using tweets. In: International Conference on Power, Control, Signals and Instrumentation Engineering, ICPCSI 2017, pp. 2734–2739 (2018). https://doi.org/10.1109/ICPCSI.2017.8392217
9. Boulos, M.N.K., Sanfilippo, A.P., Corley, C.D., Wheeler, S.: Social web mining and exploitation for serious applications: technosocial predictive analytics and related technologies for public health, environmental and national security surveillance. Comput. Methods Programs Biomed. **100**, 16–23 (2010). https://doi.org/10.1016/j.cmpb.2010.02.007

10. de Carvalho, V.D.H., Costa, A.P.C.S.: Social web mining as a tool to support public security sentiment analysis. In: Freitas, P.S., Dargam, F., Ribeiro, R., et al. (eds.) 5th International Conference on Decision Support System Technology, pp. 164–169. EURO Working Group on Decision Support Systems, Funchal (2019)
11. Gerber, M.S.: Predicting crime using Twitter and kernel density estimation. Decis. Support Syst. **61**, 115–125 (2014). https://doi.org/10.1016/j.dss.2014.02.003
12. Nepomuceno, T.C.C., Costa, A.P.C.S.: Spatial visualization on patterns of disaggregate robberies. Oper. Res. (2019). https://doi.org/10.1007/s12351-019-00479-z
13. Pereira, D.V.S., Mota, C.M.M., Andresen, M.A.: The homicide drop in Recife, Brazil: a study of crime concentrations and spatial patterns. Homicide Stud. **21**, 21–38 (2017). https://doi.org/10.1177/1088767916634405
14. Henriques de Gusmão, A.P., Aragão Pereira, R.M., Silva, M.M., da Costa Borba, B.F.: The use of a decision support system to aid a location problem regarding a public security facility. In: Freitas, P.S.A., Dargam, F., Moreno, J.M. (eds.) EmC-ICDSST 2019. LNBIP, vol. 348, pp. 15–27. Springer, Cham (2019). https://doi.org/10.1007/978-3-030-18819-1_2
15. Pang, B., Lee, L.: Opinion mining and sentiment analysis. Found. Trends Inf. Retr. **2**, 1–135 (2008). https://doi.org/10.1561/1500000011
16. Kharrat, S., Kchaou, S.: Lexicon-based methods for sentiment analysis. Comput. Linguist. **37**, 267–307 (2007)
17. Brum, H.B., Das Graças Volpe Nunes, M.: Building a sentiment corpus of tweets in Brazilian Portuguese. In: LREC 2018 - 11th International Conference on Language Resources and Evaluation, pp. 4167–4172 (2019)
18. Chathuranga, J., Ediriweera, S., Hasantha, R., et al.: Annotating opinions and opinion targets in student course feedback. In: LREC 2018 - 11th International Conference on Language Resources and Evaluation, pp. 2684–2688 (2019)
19. Turchi, M., Negri, M.: Automatic annotation of machine translation datasets with binary quality judgements. In: Proceedings of the 9th International Conference on Language Resources and Evaluation, LREC 2014, pp. 1788–1792 (2014)
20. Win, S.S.M., Aung, T.N.: Automated text annotation for social media data during natural disasters. Adv. Sci. Technol. Eng. Syst. **3**, 119–127 (2018). https://doi.org/10.25046/aj030214
21. Walkowiak, T., Gniewkowski, M.: Distance measures for clustering of documents in a topic space. Adv. Intell. Syst. Comput. **987**, 544–552 (2020). https://doi.org/10.1007/978-3-030-19501-4_54
22. Cook, P., Brinton, L.J.: Building and evaluating web corpora representing national varieties of English. Lang. Resour. Eval. **51**, 643–662 (2017). https://doi.org/10.1007/s10579-016-9378-z
23. Hovy, E., Lavid, J.: Towards a 'science' of corpus annotation: a new methodological challenge for corpus linguistics. Int. J. Transl. **22**, 13–36 (2010)
24. Baccouche, A., Garcia-Zapirain, B., Elmaghraby, A.: Annotation technique for health-related tweets sentiment analysis. In: 2018 IEEE International Symposium on Signal Processing and Information Technology, ISSPIT 2018, pp. 382–387 (2019). https://doi.org/10.1109/ISSPIT.2018.8642685
25. Zhang, H., Gan, W., Jiang, B.: Machine learning and lexicon based methods for sentiment classification: a survey. In: 2014 11th Web Information System and Application Conference (WISA). IEEE, New York, NY, USA, pp 262–265 (2014)
26. Neogi, P.P.G., Das, A.K., Goswami, S., Mustafi, J.: Topic modeling for text classification. In: Mandal, J.K., Bhattacharya, D. (eds.) Emerging Technology in Modelling and Graphics. AISC, vol. 937, pp. 395–407. Springer, Singapore (2020). https://doi.org/10.1007/978-981-13-7403-6_36
27. Dahal, B., Kumar, S.A.P., Li, Z.: Topic modeling and sentiment analysis of global climate change tweets. Soc. Netw. Anal. Min. **9**, 1–20 (2019). https://doi.org/10.1007/s13278-019-0568-8

28. Cunningham-Nelson, S., Baktashmotlagh, M., Boles, W.: Visualizing student opinion through text analysis. IEEE Trans. Educ. **62**, 305–311 (2019). https://doi.org/10.1109/TE.2019.2924385

29. Groß-Klußmann, A., König, S., Ebner, M.: Buzzwords build momentum: global financial twitter sentiment and the aggregate stock market. Expert Syst. Appl. **136**, 171–186 (2019). https://doi.org/10.1016/j.eswa.2019.06.027

30. Srinivasan, B., Mohan Kumar, K.: Flock the similar users of twitter by using latent Dirichlet allocation. Int. J. Sci. Technol. Res. **8**, 1421–1425 (2019)

31. Aggarwal, C.C.: Machine learning for text. Springer, Cham (2018). https://doi.org/10.1007/978-3-319-73531-3

32. Blei, D., Carin, L., Dunson, D.: Probabilistic topic models. IEEE Signal Process. Mag. **27**, 55–65 (2010). https://doi.org/10.1109/MSP.2010.938079

33. Ravi, K., Ravi, V.: A survey on opinion mining and sentiment analysis: tasks, approaches and applications. Knowl.-Based Syst. **89**, 14–46 (2015). https://doi.org/10.1016/j.knosys.2015.06.015

34. Yang, P., Chen, Y.: A survey on sentiment analysis by using machine learning methods. In: 2nd Information Technology, Networking, Electronic and Automation Control Conference (ITNEC), pp 117–121. IEEE (2017)

35. Asghar, M.Z., Kundi, F.M., Ahmad, S., et al.: T-SAF: Twitter sentiment analysis framework using a hybrid classification scheme. Expert Syst. **35**, 1–19 (2018). https://doi.org/10.1111/exsy.12233

36. Khan, F.H., Bashir, S., Qamar, U.: TOM: Twitter opinion mining framework using hybrid classification scheme. Decis. Support Syst. **57**, 245–257 (2014). https://doi.org/10.1016/j.dss.2013.09.004

37. De Arruda, G.D., Roman, N.T., Monteiro, A.M.: An Annotated Corpus for Sentiment Analysis in Political News, pp. 101–110 (2015)

38. dos Santos, H.D.P., Woloszyn, V., Vieira, R., Blogset, B.R.: A Brazilian Portuguese blog corpus. In: LREC 2018 11th International Conference on Language Resources and Evaluation, pp. 661–664 (2019)

39. Freitas, C., Motta, E., Milidiú, R.L., César, J.: Sparkling Vampire... LOL! Annotating opinions in a book review corpus. In: Aluísio, S., Tagnin, S.E.O. (eds.) New Language Technologies and Linguistic Research: A Two-Way Road, pp. 128–146. Cambridge Scholars Publishing, Newcastle upon Tyne (2013)

40. de Souza, K.F., Pereira, M.H.R., Dalip, D.H.: UniLex: Método Léxico para Análise de Sentimentos Textuais sobre Conteúdo de Tweets em Português Brasileiro. Abakós **5**, 79 (2017). https://doi.org/10.5752/p.2316-9451.2017v5n2p79

41. Rosa, R.L., Rodriguez, D.Z., Bressan, G.: SentiMeter-Br: A new social web analysis metric to discover consumers' sentiment. In: Proceedings of the International Symposium Consumer Electronics, ISCE, pp. 153–154 (2013). https://doi.org/10.1109/ISCE.2013.6570158

42. Bird, S., Klein, E., Loper, E.: Natural Language Processing with Python. O'Reilly Media Inc., Sebastopol (2009). https://www.nltk.org/

43. Reinoso, G., Farooq, B., Forum, C.T.R.: Urban pulse analysis using big data. In: Canadian Transportation Research Forum 50th Annual Conference. Transportation Association of Canada (TAC), Montreal, p. 16 (2015)

44. Blei, D.M., Ng, A.Y., Jordan, M.I.: Latent Dirichlet allocation. J. Mach. Learn. Res. **3**, 993–1022 (2003)

An E-government Procurement Decision Support System Model for Public Private Partnership Projects in Egypt

Karim Soliman[1]($\boxtimes$) and Nada El-Barkouky[2]

[1] Arab Academy for Science, Technology, and Maritime Transport, Cairo, Egypt
karim.mohamed@aast.edu
[2] Ministry of Transport, Cairo, Egypt
nada.elbarkouky@mot.com.eg

Abstract. The use of Decision Support Systems (DSS) has become a successful approach over the past years as it relies on the use of knowledge rather than information. Moreover, the tremendous increase in communication and information technology applications is encouraging decision-makers to incorporate the extensive analysis of previous transactions and historical experiences, along with the computer applications, to take the critical decisions related to procurement. However, there are several challenges that face the procurement decision-making process in any organization, especially, the governmental entities, such as how to choose the best suppliers to deal with. In this sense, this paper aims to present the impact of using DSS to help facilitate a better e-procurement cycle, which will involve better decision-making in the selection of the most convenient suppliers based on the previous historical record on the nominated suppliers. To achieve this aim, a DSS tool that uses decision trees named Expert System Builder (ES-builder) has been used. This paper will be specific to a case study on the new Public-Private Partnership (PPP) law number (#) 67 in Egypt, which is a law concerned with the cycle that any governmental entity engaging in projects with the private sector.

Keywords: E-government procurement · Private Public Partnership (PPP) ·
Decision Support Systems (DSS) · Egypt · ES-builder

1 Introduction

Sharing information, communication, and trust between different entities involved in the supply chain (SC) have a great impact on the performance of the entire chain [1]. Moreover, the increased competition between companies entails them to implement several electronic data interchange (EDI) technologies and further using e-procurement [1, 2]. Several pieces of research [3, 4] revealed the rising trend of applying the e-procurement in industry and governments. However, the government sectors are far behind the industry [3]. [5–7] declared that e-procurement has many successful applied initiatives that support and manage public procurement activities with higher transparency. The development of e-procurement processes has shaped several new methods and options that

© Springer Nature Switzerland AG 2020
J. M. Moreno-Jiménez et al. (Eds.): ICDSST 2020, LNBIP 384, pp. 112–124, 2020.
https://doi.org/10.1007/978-3-030-46224-6_9

support the procurement decision-making process in the government sector as it realizes potential savings and efficiencies that could be attained [8].

Despite all recognized gains from implementing such technology, the concept has not gained much attention in the developing countries, specifically, in the governmental sector. [9] found that the implementation efforts of e-government projects have resulted in a 60 to 85% failure rate. The reasons for this failure could be the differing contexts of the regulatory frameworks of each country [9]. Additionally, [10] mentioned that the sociocultural backgrounds, along with the political environment, may also impact the implementation of any technological efforts undertaken by the government. Another limitation of the e-government researchers conducted is that the literature focused on adopting the strategies and processes from developed countries' perspective instead of underlining the differences between the adopted strategies with the current environment [9]. Furthermore, [11] added that several challenges are facing developing countries within their public administration while trying to catch-up with the developed countries.

Thus, this research will review several e-procurement strategies adopted in the government sector and how the e-procurement models can help develop performance in the government sector. Furthermore, this research develops a decision tree model using the ES-builder tool, to present law #67, year 2010, which is concerned with the engagement of government entities with the private sector in infrastructure projects under the name PPP projects.

2 Literature Review

Procurement is considered one of the SC management activities that influence any organization's performance [1, 8]. Nonetheless, the focus on the adoption of online processes, such as e-government and e-procurement, has been recognized by many researchers [9, 10, 12]. Accordingly, the term "e-government procurement" has become an important matter of discussion for both academics and practitioners [13]. Likewise, a governmental procurement process has similar features with that of a private sector's company since both share a single objective: to find alternative sources of suppliers with the cheapest price and reasonable quality [14]. Several pieces of research reported the numerous benefits revealed through the consideration and adoption of the PPP instruments [9, 10]. Additionally, [7] highlighted the great opportunities gained from the PPP adoption, specifically for developing countries, as they allow governments to manage their assets efficiently while focusing on the service quality of the citizens [4]. However, this cannot be attained without increasing the governmental institutional capacities with and an affordable technological enhancement [15].

From this perspective, this section discusses and evaluates the foundation of the e-procurement concept from the traditional procurement concept. Then, various e-government procurement pros and cons will be highlighted. Further, the concept of e-government procurement in developing countries will be investigated. Finally, the PPP concept, benefits and e-governmental rationale in developing countries will be deliberated. A systematic review process was adopted to gain deep understanding of the research topic. In the first step, the Egyptian Knowledge Bank (EKB) was used as the main search engine, which enabled the access to a diversity of foremost business and management databases, including EBSCOhost, Emerald, ScienceDirect, World Bank and IEEE. Additionally, the EKB granted access to several governmental periodicals and reports. The key words used in searching are: e-government, e-procurement, PPP, DSS.

2.1 E-procurement Concept, Benefits, and Challenges

The importance of procurement became significant during the industrial revolution when the strategic role of procurement to get the railway materials were considered in Marshall Kirkman's book [16]. Despite the importance of procurement to any organization, [17] stated that procurement is a cost center function, where unnecessary paperwork can lead to many errors. Moreover, he added that an extra 50% profit can be attained by reducing the purchasing cost with only 10%. It is worth mentioning that in the late 1960s, competitive bidding and contract management started to gain more popularity [18]. However, the high-paced competitive environment has made the procurement function more complex facing several challenges due to the strategic importance of this function [19–21].

The e-procurement fits most of the features required by the modern business, through the inter-organizational information system, or any mean of electronic data interchange (EDI) [7]. The procurement process, before 1997, was limited to a few well-established organizations using EDI [12]. Nevertheless, the concept of e-procurement started to gain widespread adoption by many private and public sector organizations. According to [9, 14], the e-procurement entails all the activities related to the acquisition of a good or a service to the organization, such as selection of the products, purchase requisition, management authorization, ordering and delivering of the products, proceeding with the payment, and finally, monitoring the suppliers' performance.

Several studies have been published on e-procurement, which has not only advanced the research area but also identified several dimensions of benefits and challenges related to e-procurement application. The following table (Table 1) summarizes the most related benefits and challenges on e-procurement implementation.

Table 1. Summary of e-procurement benefits and challenges.

E-procurement benefits	E-procurement challenges
Increasing efficiency & effectiveness • Easier coordination and integration of procurement activities • Increasing accessibility • Online contracts (exchange and reverse auctions) • Reducing any human borne error • Electronic auditing and monitoring tools • Increase competitiveness and market share • Enhancing sustainability Source [11–14]	Cultural and legislative barriers • Unskilled human resource • Trust issues • Unwillingness to share confidential information • Resistance to change • Poor commitment • The threat of forward integration of suppliers • Lack of capacity to share risk • Lack of shared goals and objectives • Incompatible metrics • Inadequacies in government policies Source [11, 12, 17]

Accordingly, many countries across the globe have considered the implementation of the e-government procurement concept to handle any public procurement process, including the acquisition process of resources, especially maintenance, repair, and operating (MRO) items, and the increasingly required components and supplies [6, 7, 17]. Moreover, several authors approved that the utilization of e-procurement would lead to massive savings in the cost of public procurement while maintaining the desired level of quality [8, 9].

2.2 Adopting E-procurement in Governments

Any fund spent by any governmental sector belongs to the public society, it is the responsibility of the decision-makers in the government to reduce any economic losses while elevating the value for money [18]. The e-government Procurement is defined as *"the collaborative use of information and communications technologies by government agencies, the bidding community, regulatory and oversight agencies, other supporting service providers, and civil society in conducting ethical procurement activities in the government procurement process cycle for the procurement of goods, works, and services and the management of contracts, thereby ensuring good governance and value for money in public procurement and contributing to the socio-economic development of country"* [14]. As clear in the above definition, the e-procurement takes a different cycle/approach in the government sector than a normal e-procurement cycle in private organizations. It is good to mention that the automation of the e-government procurement via using DSS would create a considerable impact on the government performance in their e-procurement as it builds on the historical records besides knowledge built across different top management as ministries, public authorities, etc. In this context, the use of DSS would be very much valuable to capture the experiences built across previous transactions to be able to direct the latter to the most effective and efficient decisions as will be shown later in this research.

A report published by [17] summarized the benefits of e-government procurement as shown in Table 2.

Table 2. Benefits of e-government procurement.

	E-government procurement benefits	
1	Governance	Transparency, accountability, corruption control, and rule of law.
2	Effectiveness of government	Efficiency, value for money, and civil society awareness
3	Development of markets	Competitiveness, business development, and regional development

Source: [17].

Despite the previously discussed benefits of e-government procurement, some of these benefits could not be achieved without an original reform procurement strategy, which incorporates clear procedures, workflow, and monitoring. In some countries, this reform requires introducing effective management systems and controls where few currently exist, especially in the developing countries [18]. The reason behind this is the weaknesses of their institutional structure [7, 9], along with the vulnerable trust relationship between the public and civil stakeholders [15, 22].

With the introduction of the various electronic commerce web-based technologies, the use of e-procurement became feasible to all organizations [1, 14]. However, the transformation from the conventional procurement to the e-procurement requires the analysis and design of innovative processes to reduce the bureaucratic circulation of the documents, along with a clear and distinct description of the processes [23].

2.3   Public–Private Partnerships: Concept, Benefits, and E-government Rationale

The expression PPP covers a range of different arrangements that involve the use of private sector funding and expertise [15]. Furthermore, the PPPs facilitate the creation of urgently needed public infrastructure, including the delivery of roads, rail, water, and sewage treatment, hospitals, schools, etc. [24]. Additionally, PPPs also allow government departments to cope with the growing demand for high-quality public services by improving or updating existing infrastructure [9, 19]. The public partner is represented by the government at a local, state and/or national level. The private partner can be a privately-owned business, public corporation or consortium of businesses with a specific area of expertise.

The PPP concept consists of wide-ranging relationships between the private and public sectors to improve both and existing or new public service [14, 22]. From the government point of view, adopting a PPP model would help government bodies to tap into the innovative flexibility and state-of-art technologies provided by the private sector while gaining from their expertise to stimulate the personnel government sector efficiency [15, 22, 24]. On the other side, a PPP project is considered a novel profit stream to the private sector by delivering a more extensive range of services to an otherwise restricted market [15].

The different models of PPP funding are characterized by which partner will take responsibility for owing and preserving the governmental assets at the different stages of the project [24], such as Design-Build (DB), Operation and Maintenance (O&M), Design-Build-Finance-Operate (DBFO), Build-Own-Operate (BDO), Buy-Build-Operate (BBO), Build-Lease-Operate-Transfer (BLOT), Operation License, and Finance only. Subsequently, the critical PPP success factors and enabling structures, such as e-government procurement, has gained increasing attention of study [15, 17, 19]. However, [17] affirmed that evidence from previous researches lacked the structured tools that support the early planning and decision-making steps while taking the different perceptions of both the PPPs and e-government into consideration. Hereafter, there is an urge to find a tool that helps decision-makers in the governments to self-assess the project circumstances when being engaged in a PPP project. For this reason, this paper will contribute to building an online model that can government bodies when engaging in PPP projects, which typically discusses the laws, rules, and regulations that must be applied/followed when engaging in PPP projects. Based on the above-mentioned benefits, PPPs and e-government have been proven as fruitful mechanisms for economic growth. However, practitioners face a variety of challenges and obstacles when trying to implement PPPs projects in the developing countries. Thus, in the upcoming section, the status of e-government procurement in Egypt will be highlighted. Moreover, a brief description will be underlined on the law that sets the rules and regulations for getting engaged in PPP projects (law #67), and how can decision-makers use can incorporate the role of information and communications technology (ICT) and DSS to facilitate the e-government procurement decision-making process.

3 The Concept of E-government Procurement in Egypt

3.1 Background of the Case

In Egypt, government authorities abide by law #67 year 2010 when undergoing infrastructure projects with the private sector. The Ministry of Finance (MOF) issued this law to regulate and outline the conditions that shall be followed by government authorities through the lifecycle of the project.

[2–5] indicated that it is of great benefit to the developing countries to move towards e-government as it can bridge the gaps and enhance services the government provides to citizens aiming to minimize usual governmental procedures. Furthermore, [7, 8] explained that for government to develop a better relationship with citizens they need to work on providing all sources of information to citizens in order to answer all their needs and concerns whereas, [16] has also debated that information is one of the biggest factors that contribute to better decision making. Moreover, [18, 22] have also highlighted that although e-government could be a source for providing citizens information they are looking for, still citizens need support and guidance to understand governmental information.

On the other hand, [24] indicated that before listing the procedures and the rules that must be considered for the submission, conclusion, and implementation of participation contracts, the following introductory remarks must be taken as guidelines for engaging in PPP projects:

- The partnership contract with the private sector is a contract entered into by the administrative body with the desired company. The project, which is committed to the establishment of the winning bidder in the competition contract participation and the sole purpose of the company, shall be to carry out the work or works provided in the Participation contract.
- It is not a form of participation governed by the law regulating private sector participation. in infrastructure projects, services, and utilities. Moreover, the administrative body and the private sector is prohibited to implement a joint venture, which is what comes out of the scope of application of the Egyptian Law #67 of 2010 and its implementing regulations.
- The participation contract may not be less than five years and not more than thirty years from the date of completion of construction works or completion of development works, and the value of the contract shall not be less than one hundred million Egyptian pounds.
- The partnership contract with the private sector is governed by a set of rules and procedures contained in:

- The law regulating private sector participation in infrastructure projects in public services and facilities promulgated by Egyptian law #67 of 2010.
- The executive regulations of the aforesaid law issued by the decision of the Chairman of the Council Ministers #238 of 2011.
- Other provisions stipulated in the partnership contract concluded between the Administrative entity and Project Company.
- The provisions of the Public Liability Laws shall not apply to contracts of participation with the private sector.
- The provisions of the tenders and auctions Law shall not apply to such contracts of Egyptian law #89 of 1998 and its implementing regulations.

The main stages of the Egyptian law #67 for the year 2010 for the PPP projects are the study stage and the approval of the higher committee for participation, the pre-tendering stage, the tendering and awarding stage and, the implementation stage.

3.2   Using DSS in PPP E-government Procurement Process

Decision support systems aids in the decision-making process, unlike the usual information systems. Traditional information systems mainly focus on processing data into information as explained by [24, 25]. On the other hand, [23] compared the evolution in information to the evolution in the industrial revolution where analysis and processing take place on information rather than just data. In this respect, processing results contributes to better decision-making based on the analysis of information rather than data [26].

Additionally, [24, 27, 28] indicated several approaches to design a decision support system varying from if-then rules, decision trees, linear programming, or fuzzy logic. In this paper, the researchers aimed to select one of the simplest forms of DSS, which is decision trees.

[23] introduced decision trees as a structured approach for problem-solving to support decision-making based on building all possible course of action that contributes to decision making. Decision trees are simple because they are presented in the form of a question and answer, so a user will likely jump to a conclusion starting from a simple question drilling down to a conclusion based on the relevant branch of the tree.

In order to apply a decision tree model, [23 28] applied Expert System (ES-builder) as a software program that contributes to building decision trees. ES-builder is a free desktop program that builds decision tree models based on three main pillars (A-V-C), A, indicates attributes which is the question; V, indicates the value which is the choices relevant to the question; C, indicates conclusion which is the final decision that comes out through the tree branch the user will be taking. One of the benefits of the ES-builder tool that will help users is to see the output in different forms. The tool can display the model structure in either question and answers view, decision tree view, a knowledge base view, or a decision table view. The most dominant and user-friendly view that users can follow is the questions and answers view, where users keep answering questions, until getting to the conclusion/decision based on the answers they provide as shown in Figs. 1, 2, 3, 4, 5, 6, 7 and 8, in the next section. Figure 1, shows the main screen where upon pressing search, the questions and answers view appears. Figure 2, shows sample questions that appear to the users where they keep going forward through the questions until reaching a decision relevant to the answers provided. Figures 3, 4, 5, and

6 represents a sample of the decision tree. Figure 7 represents another form of output which is the knowledge base, it represents the content of the whole tree/model in the form of an If-then-else until reaching the conclusion. Finally, the last output form ES-builder can generate is, the attributes list as shown in Fig. 8, which is a tabular form that shows the content of the model in a tabular view.

In the light of the information presented above, the selected ES-builder tool will build a decision tree model as a framework for conducting E-procurement in the government sector in PPP projects, based on the laws and regulations that shall be followed to engage in PPP projects as a governmental body.

3.3 E-procurement Model Development Under PPP in Egypt

This model is built based on law #67, year 2010 and its executive regulations based on the chairman of the council Ministers decree #238, year 2011. The mentioned law was translated and transformed into a decision tree model to support government entities undergoing a PPP project. The Model explains the major pillars/guidelines that control and aids these entities knowing if it is possible to proceed in engaging in this project in addition to, the conditions that shall be followed in law #67 for PPP Projects. The model goes through 4 major stages 1- The study stage and higher committee for PPP stage, 2- The Pre-tendering phase, 3- The Tendering Phase, 4- The implementation phase. The 4 stages goes through a number of questions where users keep going forward with the questions if the answer is (Yes), user proceeds with the next questions until ensuring all stages required to engage in PPP are fulfilled. If the answer is (No) the flow of the questions stops at this stage, and a decision is given to the user that "You cannot engage in PPP" because the flow of the model is dependant, which means users can't proceed if any of the stages was given a (No) answer as it indicates that requirements to engage in PPP projects is not fulfilled as per the Law #67 year 2010. In case the users gets a (No) and aims to solve the problem to be able to engage in PPP, users will have to track back where they ended in the tree, and try to complete the requested document/steps that shall be taken in order to proceed with the flow of the tree.

The following section elaborates how the model can be followed to reach an optimum decision based on answering the tree questions and getting conclusions based on the inserted information from the answered questions which takes the entity using this model to a decision whether it is possible for their entity to engage in a PPP project or not.

The researchers have validated the model structure via 15 interviews with different stakeholders from the government sector. The ministry of transport, being one of the ministries that engage in infrastructure projects; the following authorities which are affiliated to the ministry of transport were interviewed (Egyptian national railways, Damietta port, Alexandria port, national authority for tunnels, river transport authority, Land and dry ports) and the Ministry of Housing. Two interviewees where interviewed to give their input about the model. Feedback was highly positive as the major findings was that they all engage in different infrastructure projects, and following the PPP law is a must. Therefore, the model is a good approach to facilitate the procurement steps that need to be followed in order to proceed with a PPP project.

As seen in Fig. 1 below, the model begins with a home screen, where the user can start by clicking search to get to the questions and answers view, which is as mentioned in the previous section the most dominant interface the users can rely on. Then, users move to the first question where users click on the relevant choice as shown in Fig. 2.

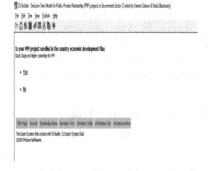

Fig. 1. The main screen   Fig. 2. Questions-answers view

Based on the above approach the cycle starts with a group of questions (included in Sect. 1 of the tree) covering the "Study Stage and Higher committee for PPP". Figure 3 represents this stage, where it is composed of 10 questions, which starts by enrolling the targeted PPP project in the country's economic development plan and ends at sharing an introductory memo with PPP unit, the ministry of finance.

Section 2 of the tree covers 5 questions on the "Pre-tendering phase", as shown in Fig. 4 below. This stage covers all needed preparations for tendering ending at approval from PPP unit, ministry of finance for pre-qualifications. Section 3 is the "Tendering Phase" which is composed of 17 questions as shown in Fig. 5 below. The scope of this stage starts with arranging committee for pre-qualifications ending at setting the

Fig. 3. Section 1 of the tree "The study stage and higher committee for PPP"

Fig. 4. Section 2 of the tree "The pre-tendering phase

estimated project values and submitting these values to the PPP unit, the ministry of finance for a government comparative. Section 4 of the tree is the "Implementation Phase" which is composed of 3 questions as shown in Fig. 6 below. This stage ensures that the targeted PPP project has multiple bidders and not a single offer, because in case only 1 bidder applied for the project the decision is "cannot engage in PPP" since the mentioned law #67, year 2010 prohibits single bidder.

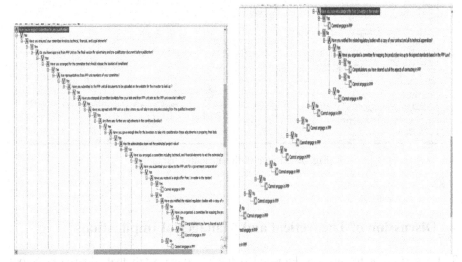

Fig. 5. Section 3 of the tree "The tendering phase"

Fig. 6. Section 4 of the tree "The implementation phase"

ES-builder can also show output in a knowledge base form, where it displays the model in if – then – else form. The program drills down through all the tree branches and take you through all the possible course of action without having to navigate through questions, and finally shows the result or conclusion in an "if – then" statement as shown in Fig. 7.

Fig. 7. Sample of the knowledge base view

Furthermore, the ES-builder can create an attribute list as shown in Fig. 8 below, where it displays the breakdown of the model in an attributes tabular form.

Attribute (Decision Tree)	Attribute (Expert System)	Values	Image	Help Notes
Is your PPP project enrolled in the country economic development Plan	Is your PPP project enrolled in the country economic development Plan	• Yes • No		Study Stage and Higher committee for PPP
Have you prepared the introductory document	Have you prepared the introductory document	• Yes • No		Study Stage and Higher committee for PPP
Have you explained project scope and identified the duration?	Have you explained project scope and identified the duration?	• Yes • NO		Study Stage and Higher committee for PPP
Have you Identified the administrative body that is tendering the project and all other administrative bodies relevant to the project	Have you Identified the administrative body that is tendering the project and all other administrative bodies relevant to the project	• Yes • No		Study Stage and Higher committee for PPP
Have you identified the role of the administrative body and defined the obligations?	Have you identified the role of the administrative body and defined the obligations?	• Yes • No		Study Stage and Higher committee for PPP

Fig. 8. Sample of the attributes list table

4   Discussion of Theoretical and Managerial Implications

To the best of the researchers' knowledge, this is the first study that adopts a DSS tool to conduct an e-government procurement to PPP projects in Egypt. As mentioned in the previous sections, [26] debated that e-government solutions are becoming a valuable source of information for citizens. In addition [23] acknowledged that although e-government could be a source for providing citizens information they are looking for, still citizens need support and guidance to understand governmental information. In this context, it comes to be effective to create a decision tree model using ES-builder, as a DSS tool, that will help decision-makers to decide on whether it is applicable to engage in PPP projects following the conditions/instructions of Egyptian law #67 for year 2010. Besides the theoretical contribution discussed, this research can act, as a guiding model for any governmental entity engaging in PPP projects, as well as it will clarify the phases in a systematic approach showing the guidelines of the mentioned PPP law.

5   Conclusion, Recommendations, and Limitations

This paper assists government entities engaging in PPP projects by considering all the conditions of law #67 for the year 2010. The model acts as a guideline for the users who aim to engage in PPP projects with the government sector. Law #67 year 2010, PPP unit in the ministry of Finance, is the regulator for PPP projects in the government they prohibit missing any of the mentioned stages in the model, and if the users missed any of the steps guided through the model, the company is dismissed from the tender. So, the model adds value to private sector companies in two ways; 1- it could give the users who aim to engage in PPP projects guidelines for what documents would they need and what stages will be followed throughout the phases of PPP. 2- It assists the users who are not much acquainted with the mentioned law to decide whether they could engage in PPP projects or not based on a facilitated question/answer format through the ES-builder tool. The paper takes a systematic approach by going through all the different stages of the law starting from the study stage until the implementation stage. A decision tree approach is used to facilitate all the conditions of the law, which is a friendly approach for the users dealing with the government entities. This model is recommended for any government entity undergoing infrastructure projects where they must follow the law #67, year 2010 in order to proceed to contract with a private sector company. The model makes it easier for the users to know whether they are allowed or prohibited without having to spend much time exploring the conditions set by the mentioned law. This research is limited to only government entities undergoing infrastructure projects with private sector companies and future work can be further made on other laws that regulate procurement in the government as it would assist decision-makers in making the right decisions via using this decision tree model.

References

1. Hsin Chang, H., Tsai, Y.C., Hsu, C.H.: E-procurement and supply chain performance. Supply Chain Manage. Int. J. **18**(1), 34–51 (2013)
2. Witjes, S., Lozano, R.: Towards a more Circular Economy: proposing a framework linking sustainable public procurement and sustainable business models. Resour. Conserv. Recycle. **112**, 37–44 (2016)
3. Osei-Kojo, A.: E-government and public service quality in Ghana. J. Publ. Aff. **17**, e1620 (2017)
4. Mates, P., Lechner, T., Rieger, P., Pěkná, J.: Towards e-Government project assessment: European approach. Zbornik radova Ekonomskog fakulteta u Rijeci: časopis za ekonomsku teoriju i praksu **31**(1), 103–125 (2013)
5. Aladwani, A.M.: Corruption as a source of e-Government projects failure in developing countries: a theoretical exposition. Int. J. Inf. Manage. **36**(1), 105–112 (2016)
6. Potnis, D.D.: Measuring e-Governance as an innovation in the public sector. Gov. Inf. Q. **27**, 41–48 (2010)
7. Palaco, I., Park, M.J., Kim, S.K., Rho, J.J.: Public–private partnerships for e-government in developing countries: an early stage assessment framework. Eval. Program Plann. **72**, 205–218 (2019)
8. Martins, J., Veiga, L.: Electronic government and the ease of doing business. In: Proceedings of the 11th International Conference on Theory and Practice of Electronic Governance, pp. 584–587. ACM (2018)
9. Basílio, M.: The degree of private participation in PPPs: evidence from developing and emerging economies. In: The Emerald Handbook of Public–Private Partnerships in Developing and Emerging Economies: Perspectives on Public Policy, Entrepreneurship and Poverty, pp. 81–111). Emerald Publishing Limited, Bingley (2017)
10. Ferk, B., Ferk, P.: Top 10 reasons why (Not) and how (Not) to implement PPPs in the developing and emerging economies. In: The Emerald Handbook of Public-Private Partnerships in Developing and Emerging Economies: Perspectives on Public Policy, Entrepreneurship and Poverty, pp. 3–44. Emerald Publishing, Bingley (2017)
11. Altayyar, A.: Investigating e-procurement barriers within six Saudi Arabian SMEs (2017)

12. Birou, L., Lutz, H., Zsidisin, G.A.: Current state of the art and science: a survey of purchasing and supply management courses and teaching approaches. Int. J. Procurement Manage. **9**(1), 71–85 (2016)
13. Kapferer, J.N., Michaut, A.: Luxury counterfeit purchasing: The collateral effect of luxury brands' trading down policy. J. Brand Strategy **3**(1), 59–70 (2014)
14. Nurmandi, A.: What is the status of Indonesia's e-procurement? Jurnal Studi Pemerintahan **4**(2) (2013)
15. Shen, L., Tam, V., Gan, L., Ye, K., Zhao, Z.: Improving sustainability performance for public-private-partnership (PPP) projects. Sustainability **8**(3), 289 (2016)
16. Wahed, M.E.S., El Gohary, E.M.: The future vision for the design of E-government in Egypt. Int. J. Comput. Sci. Issues (IJCSI) **10**(3), 292 (2013)
17. Shen, L., Zhang, Z., Long, Z.: Significant barriers to green procurement in real estate development. Resour. Conserv. Recycl. **116**, 160–168 (2017)
18. World Bank Group: Benchmarking Public-Private Partnerships Procurement 2017. World Bank, Washington DC (2017)
19. Tunji-Olayeni, P.F., Emetere, M., Afolabi, A.O.: Multilayer perceptron network model for construction material procurement in fast developing cities. Int. J. Civ. Eng. Technol. (IJCIET) **8**(5), 1468–1475 (2017)
20. Suri, P.K.: Introduction to E-governance. In: Strategic Planning and Implementation of E-Governance. Flexible Systems Management, pp. 1–24. Springer, Singapore (2017). https://doi.org/10.1007/978-981-10-2176-3_1
21. Taher, M., Yang, Z., Kankanhalli, A.: Public-private partnerships in E-government: insights from Singapore cases. In: PACIS, p. 116 (2012)
22. Klischewski, R., Askar, E.: Linking service development methods to interoperability governance: the case of Egypt. Gov. Inf. Q. **29**, S22–S32 (2012)
23. Matthew, O.O., Buckley, K., Garvey, M., Moreton, R.: Multi-tenant database framework validation and implementation into an expert system. Int. J. Adv. Stud. Comput. Sci. Eng. (IJASCSE) **5**(8), 13–21 (2016)
24. Bonczek, R.H., Holsapple, C.W., Whinston, A.B.: Foundations of Decision Support Systems. Academic Press, Cambridge (2014)
25. Araujo, L.G., Piña, A.B.S., Aidar, L.A.G., Coelho, G.O., Carvalho, M.T.M.: Recommendations and guidelines for implementing PPP projects. Built Environ. Project Asset Manage. **9**, 262–276 (2019)
26. Badasyan, N.: Project feasibility analysis economic model for private investments in the renewable energy sector. Built Environ. Project Asset Manage. **8**(2), 215–230 (2018)
27. Livieris, I.E., Drakopoulou, K., Kotsilieris, T., Tampakas, V., Pintelas, P.: DSS-PSP - a decision support software for evaluating students' performance. In: Boracchi, G., Iliadis, L., Jayne, C., Likas, A. (eds.) EANN 2017. CCIS, vol. 744, pp. 63–74. Springer, Cham (2017). https://doi.org/10.1007/978-3-319-65172-9_6
28. McGuinness, D.: Software Programming in IPT using a Simplified AJAX Architecture (2013)

Determinants of Recommendation in the Airline Industry: An Application of Online Review Analysis

Praowpan Tansitpong[✉]

NIDA Business School, National Institute of Development Administration (NIDA),
118 Seri Thai Road, Bangkok 10240, Thailand
praowpan.tan@nida.ac.th

Abstract. This study explores key determinants of airline recommendations by integrating ratings and text comments scraped from online source. Numerical review scores are used to characterize features on how passengers decide to recommend others. Text analysis technique provides information of service attributes that differentiate positive and negative comments. Using generated frequent words visualization of Word Cloud, the results suggested that positive recommenders are satisfied with human dimensions such as personality and friendly services, while negative comments suggested frequent complaints on poor operational dimensions such as on-time performance and seat comfort.

Keywords: Web scraping · Text mining · Airline recommendation · Online review · WordCloud

1 Introduction

In the past decades, airline services were unpleasant among passengers due to cost cutting strategy and no-frills operations. In order to survive in this competition, academic researchers are seeking new benchmark for desirable service characteristics that could help the airlines retain customer value and improve satisfaction. In order to improve service quality, a series of studies were conducted to determine relationships between airline service dimensions such as tangibles, schedules, services provided by ground staffs, seat comfort, on satisfaction [1–4]. These literatures suggested that recommendation is very useful for airline to understand the likelihood of customer referring and influencing others to participate with the service from their previous experiences [5, 6]. These studies investigated the influence of recommendation by determining how customer rates and recommends the airline service to others. As neglected by literature in marketing and operations management, useful information from recommendation scores can be applied to predict service outcome from reviews and comments. Online reviews and comments can also be used for both quantitative and qualitative purposes. This study will utilize recommendations by combining both using quantitative and qualitative analytical tools to examine desirable service attributes of airlines that could improve the service for passengers.

© Springer Nature Switzerland AG 2020
J. M. Moreno-Jiménez et al. (Eds.): ICDSST 2020, LNBIP 384, pp. 125–135, 2020.
https://doi.org/10.1007/978-3-030-46224-6_10

Social media and online reviews allow airline passengers to voice out their comments through websites. The online reviews give feedback to the airline industry for improving on dimensions that have been lacking in their current operations. Their services are complicated; the airline industry provides services with both tangible (such as seat comfort, meals, magazines, etc.) and intangible (such as flight attendant efficiency, responding to request, attitude, etc.) aspects [7]. Moreover, the industry is highly competitive; passengers have multiple product options and airline choices to choose from. Reviews from online sources are available for passengers to participate with standard web platform. Skytrax website (http://www.airlinequality.com), a third-party assessor based in UK provides a ranking scheme that is endorsed by all major airlines around the world, incorporates detailed airline service scores (such as ratings for check-in, transfer, arrival, seat comfort, quality of meals, lounge, washrooms, cabin cleanliness, enthusiasm and attitude of staff, friendliness and hospitality, etc.). These detailed scores are reported more than 100,000 detailed service items is published for each airline. Hence, by using available text and rating data online, this study is seeking to identify desirable aspects of service attributes that can be used to predict recommendations for airlines.

Recommendations are useful information to learn about passengers' experiences with the airlines, multiple studies argued that only recommendation score is not adequate due to variations in airline passenger expectation of service quality [5, 8]. Hence, modern literature suggested that user-generated content as self-ratings, online comments, and service reviews, are better determinants to evaluated consumer's recommendation [9–11]. Other line of marketing literature also suggested that passengers would act as promoters when choosing and recommending the products or service to others [12–14], this metric is called Net Promoter Scores (NPS). According to the previous literature, NPS has been rarely used in the airline context. Net Promoter Score is an index that measures the willingness of customers to recommend a company's products or services to others [15]. NPS is used as a proxy for gauging the customer's overall satisfaction with a company's product or service and the customer's loyalty to the service.

This paper will compare the usage of recommendation and NPS scores, by comparing their predictive accuracy. Both scores derived from customer ratings and comments could be a fruitful resource for researcher to connect a linkage between service dimensions and intention to recommend to others, the text comments also reveal qualitative attributes of the service. By utilizing a concept of recommendation, this study seeks to predict which passengers that could be influencers for recommendations and provide the connection from a usage of numerical ratings to predict and distinguish recommenders from others. This study also utilized online web reviews to build a prediction model for text comments and to characterize qualitative features on how passengers decide to recommend other.

2 Methodology

2.1 Data Collection

Airline review dataset is collected from customer reviews collected on Skytrax website (http://www.airlinereview.com). The process of data collection is automated or scraped through Python program source code. Web scraping process took place between September to October of 2017, a total of 7,000 reviews are stratified based on reviews from domestic US airlines (American Airlines, Delta Airlines, JetBlue Airways, Southwest Airlines, Spirit Airlines, US Airways, and United Airlines) with English reviews. From the reviews, top comment contains more than 19,000 words. Scores for seat comfort, cabin staff service, food and beverages, inflight entertainment, ground service, and value for money are out of 5-point, rating is out 10 points. Recommendation is binary values. For text data, the website sub-content in each page provides passengers reviews. In this airline review section, the website displays individual passenger review from airline list and top comment such as "Excellent on-board service" and a paragraph with of their detailed text comments. Other ranking for aircraft type, type of travelers, cabin flown, route, date flown, are also published on each page.

Marketing literature has applied Net Promoter Score (NPS) as a metric to measure differences between promoters (positive recommenders), passively satisfied, and detractors [5]. The key idea of the NPS is that the measurement systems of customer satisfaction and customer retention do not help achieve growth compared with the word-of-mouth metric. The word-of-mouth metric asks the question, "How likely is it that you would recommend [company X] to a friend or colleague?" The scale ranges from 0 to 10; there were some studies where 10-points scale substitutes the 11-points scale with negligible difference [10]. The researchers categorized respondents into three groups, respondents who rate 6 or less are called "detractors", those rating the scale at 7 and 8 are "passive satisfied", and respondents rating a firm 9 or 10 are "promoters". This scale is applied to further understand the extend to the preferences of customers in choosing and recommending the products or service to others [16, 17].

The online reviews from flight experiences were collected by using source codes provided in Python libraries which are called UrlLib3 to manage connections with hosts. The library BeautifulSoup4 (BS4) is an interface working with HTML and XML document standards to convert Python's csv format into manageable Excel format. *The data was then split into two sets of outcomes. The first set of data uses raw binary 'Recommended' score as predictive variable. The second set of data transforms rating scores in to three scales of NPS, then separate the predicting outcomes into 'Recommenders' and 'Non-recommenders'.* The two datasets will be analyzed to compare predictability of their accuracy from logistic regression results. The dataset with superior accuracy will be used to analyze frequency of word comments or Word Cloud, together with the inferior one, to summarize qualitative aspect of attributes for airline service (Fig. 1).

"excellent on-board service"

Nick Biskinis (United Kingdom) 2nd October 2017

☑ Verified Review | Athens to London Heathrow. This is the best direct airline from the UK to Greece. Very good value fare and excellent on-board service with pre-flight sweets and during the journey, two rounds of drinks, tea/coffee separately and a hot meal. Clean A321 both ways. Flying into Heathrow was the usual stacking before the always amazing sightseeing aerial tour of London as the aircraft made its approach. I used to fly BA on this route but a large gulf has appeared in its service provision for the outer European routes: until BA substantially improves I will continue to fly Aegean from Heathrow.

Aircraft	A321
Type Of Traveller	Solo Leisure
Cabin Flown	Economy Class
Route	Athens to London Heathrow
Date Flown	September 2017
Seat Comfort	✪✪✪✪⊗
Cabin Staff Service	✪✪✪✪✪
Food & Beverages	✪✪✪✪✪
Inflight Entertainment	✪✪✪⊗⊗
Ground Service	✪✪✪✪⊗
Value For Money	✪✪✪✪✪
Recommended	✔

Fig. 1. Example of domestic US airlines scores and comments scraped from airlinequality.com

3   Results

3.1   Descriptive Summary

A proportional percentage of recommendations from six dimensions of service scores and service attributes (Cabin Staff Service, Food & Beverage, Ground Service, for Money) are created in bar a chart of Max-min scaler to standardize the data (Fig. 2). The four highest percentage for recommenders with percentage of proportion greater than 0.8 from a scale of 1 are Value for Money, Cabin Staff Service, Ground Service, and Rating for airlines. On the other hand, majority of non-recommenders tend to give low scores on Food & Beverages, Rating, Value for Money and Inflight Entertainment in which the proportional scale lower than 0.2. The NPS scores reported similar distribution of each service attributes, but divided group of reviews into three categories (promoters, neutrals, and detractors). A descriptive summary of all variables is provided in Table 1.

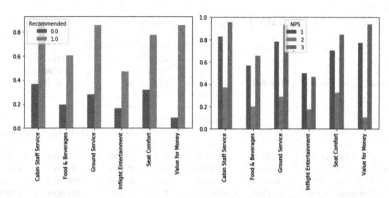

Fig. 2. Normalized scale of review scores between binary recommendations (left) and Normalized NPS scores for 'Recommenders' or Promoters and 'Non-recommenders' or Neutral and Detractors classification (right)

Table 1. Descriptive summary.

Variable	Obs	Mean	Std. Dev.	Min	Max
Recommended (derived from NPS)	7000	0.3181429	0.4657886	0	1
Raw (binary) recommendation	7000	0.277344	0.447746	0	1
Value for Money	7000	2.379571	1.597365	0	5
Cabin Staff	7000	2.595571	1.633083	0	5
Food & Beverages	7000	1.755429	1.599815	0	5
Ground Service	7000	1.381571	1.619279	0	5
Inflight Entertainment	7000	1.408714	1.62315	0	5
Seat Comfort	7000	2.338429	1.470152	0	5

3.2 A New Way to Split Dataset: 'Recommenders and Non-recommenders'

With only 27.7% of the respondents made selection of 'Recommended' to this online survey. This statistic creates opportunity for this study to understand how dimensions of service could influence behavior of recommendation. Net Promoter Score (NPS) is a metric that measures differences of respondents who are likely to promotes (positive recommenders), passively satisfied, and detractors [18–20]. The scores derived from ratings are ranges from 0 to 10, the respondents who rate 6 or less are called "detractors, or score of 1", those rating the scale at 7 and 8 are "passive satisfied, 2", and respondents rating a firm 9 or 10 are "promoters", or 3. According to new transformed 'Recommender's' scores, 31.8% of the passengers are identified as 'Recommenders', while 68.2% of the reviewers are classified as 'Non-recommenders'.

Table 2. Logistic regression results

```
Logistic regression                          Number of obs   =     7000
                                             Wald chi2(7)    =  1192.05
                                             Prob > chi2     =   0.0000
Log pseudolikelihood = -954.90643            Pseudo R2       =   0.7819
```

Recommended	Coef.	Robust Std. Err.	z	P>|z|	[95% Conf. Interval]	
ValueforMoney	2.183204	.0866291	25.20	0.000	2.013414	2.352994
CabinStaffService	.4608386	.0574479	8.02	0.000	.3482429	.5734344
FoodBeverages	.0865209	.0542116	1.60	0.110	-.0197319	.1927736
GroundService	.2083548	.0360448	5.78	0.000	.1377084	.2790013
InflightEntertainment	.1144584	.0407641	2.81	0.005	.0345623	.1943546
SeatComfort	-.0139632	.0644182	-0.22	0.828	-.1402206	.1122942
_cons	-8.998729	.315755	-28.50	0.000	-9.617598	-8.379861

3.3 Logistic Regression

Literature has utilized logistic regression technique to develop customer satisfaction model from categorical response [21–23]. A summary of logistic prediction accuracy is reported in Table 3. Accuracy of logistic regression classifier on test set is 98% and 10-fold cross validation accuracy is reported at 97.5%. By using two predictors ('Recommenders' derived from NPS and raw recommendation scores), the accuracy of 'Recommenders' model returned higher accuracy of logistic regression classifier on test set and 10-fold cross validation average accuracy. A comparison of model accuracy between using recommenders derived from NPS scores and recommendations scores are provided in Fig. 3.

Table 3. Summary of logistic regression model accuracy between recommendations derived from NPS and raw recommendation scores

	Recommenders using NPS	Raw recommendation scores
Accuracy of logistic regression classifier on test set	0.98	0.94
10-fold cross validation average accuracy	0.975	0.932

Recommenders and non-recommenders are measured by the scores derived from ratings are ranges from 0 to 10, the respondents who rate 6 or less are called "detractors, or score of 1", those rating the scale at 7 and 8 are "passive satisfied, 2", and respondents rating a firm 9 or 10 are "promoters", or 3. As the dataset of 'Recommenders' from NPS scores performed better results in logistic regression accuracy, this section will provide results from logistic regression outcome and report coefficients and P-values from regression model. Literature in the past focused on predicting airline performance [24, 25], yet

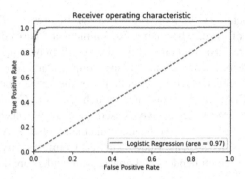

Fig. 3. ROC curve

those studies had never looked at recommendations in a context of recommendations and NPS. Logistic regression is applied to predict recommendations from 'Recommenders' by using Stata SE 11 Edition, the results suggested positive significant coefficients of Value for Money, Cabin Staff Service, Ground Service, and Inflight Entertainment, in predicting for binary response. Positive significant coefficients with P-value lower than 0.05 are *Value for Money (2.183204), Cabin staff Service (0.4608386), Ground Service (0.2083548), and Inflight Entertainment (0.1144584).*

From average ratio to total of the entire test set, 98% of service dimensions (Cabin Staff Service, Food & Beverage, Ground Service, Inflight Entertainment, Seat Comfort, Value for Money, NPS) are good predictor for Recommendation. The receiver operating characteristic (ROC) curve reports true positive and false positive rate with binary classifiers. The ROC curve confirms that recommendations is a good predictor of this model with low false positive rate (Table 3).

3.4 Text Analysis

Airline text comments are analyzed using Word Cloud. Word Cloud is Lexicon-based sentiment classification that measures the sentiment of a group of documents [26]. Lexicon-based sentiment utilizes dictionary of words (a lexicon) and each word's associated polarity score to create a group of words by using different text sizes. In order to explore comments using Word cloud, texts are gathered and split into sets of desired outcomes. In this study, the text data are split into two sets regarding the comments from recommenders and non-recommenders. The bigger and bolder the word appears, the more often words mentioned within a given text and the more important it is. By applying Natural Language Toolkit (NLTK) libraries retrieved from a website https://www.nltk.org/ [27], two visualized Word Cloud results are created from the positive and negative comments of recommenders and non-recommenders. Frequent word lists are gathered from word cloud to extract features that have been used the most in the comments.

Top comment contains more than 19,000 words. Figures 4 and 5 compare results from word cloud list of words when using 'Recommenders' derived from NPS and raw recommendation scores. The outcome of word cloud displayed significant different of words when using 'Recommenders' rather than recommendation scores. Recommenders frequently expressed 'Great', Good', 'Excellent' for the service they experienced, and mentioned 'Staff', 'Crew', and 'Attendance' which relates to Cabin Staff Service. For non-recommenders, the most frequent words used to explain the services are 'Worst', and 'Never'. Other words are found to be related to punctuality of the service such as 'Delay', 'Time', Cabin Staff Service such as 'Rude', and Seat Comfort such as 'Uncomfortable'. Frequent word lists of recommenders and non-recommenders are displayed in Table 4.

Fig. 4. Word Cloud results from 'Recommenders' vs. raw 'Recommended' classifier

Fig. 5. Word Cloud results from 'Non-recommenders' vs. raw 'Non-recommended' classifier

In a comparison between Word cloud results derived from NPS and raw recommendation classifier, frequency of words from NPS classifier contain more detailed and number of words than raw recommended one. This maybe due to word selection from positive 'Recommended' passengers tend to contain fewer comments compared to others and selection bias due to their satisfaction from the service [28]. Figures 4 and 5 display text analysis results that NPS is a better classifier for recommendation because the dataset contains more insightful comments that could help create more informative Word Clouds.

Table 4. Frequent word lists of positive and negative recommendations

Positive recommendations	Negative recommendations
Staff	Never
Great	Worst
Friendly	Fly/flying
Excellent/Good/Nice	Time
Crew	Seat
Thank	Horrible/terrible
Attendant	Delay/delayed
Efficient	Rude
Polite	Disappointed
Courteous	Uncomfortable
Service	Scam
Beyond	Tired
Warmth	Checked
Level	Downgrading

Table 5. Linkage between logistic regression results and Word Cloud

Logistic regression	Word Cloud
Predictor: recommenders	Positive word list from recommenders
Value for money	'Cost'
Cabin staff service	'Staff' 'Crew' 'Attendant' 'Polite' 'Courteous'
Ground service	'Efficient' 'Helpful'
Inflight entertainment	–
Others	'Great' 'Excellent/good/nice' 'Thank' 'Comfortable' 'Clean'

4   Conclusion

This study explores key determinants of airline recommendations from online reviews. By using online web scraping information, this study is able to distinguish recommenders from non-recommenders from online reviews. Text analysis technique provides information to characterize features on how passengers decide to recommend others and what are the attributes of service that differentiate positive and negative comments from generated Word Cloud. The results suggest that positive recommenders are satisfied with human dimensions such as personality and friendly services, while negative comments suggested frequent complaints of poor operations dimensions such as on-time performance and seat comfort.

The findings resulted from separating predicting outcome between 'Recommenders' and 'Non-recommenders' have led the investigation onto a discovery of desirable service aspects. By utilizing logistic regression to classify 'Recommenders' from passengers, the airlines can use online web reviews to help improve the understanding in detailed qualitative aspects of their current service. As confirmed by regression results, airlines should be focusing on *Cabin Staff Service, Ground Service, and Inflight Entertainment* to engage recommenders to present themselves as promoters. Passengers tend to value the cost of flying to the service they received, as shown in logistic results, it is significant to improve tangible products (such as Inflight Entertainment and Seats) and services (such as Cabin Staff Service and Ground Service) that could be seen as value for the ticket price they have paid. These service attributes should be added to new staff training in order to be competitive and to offer services with higher levels of quality (Table 5).

In addition, the results highlighted the importance of an understanding in the differences of recommendations among groups of passengers to identify new operational strategies based on the qualitative and quantitative reviews. According to the linkage to word cloud results, passengers tend to *value 'cost', 'courtesy' and 'politeness' from Cabin Staff Service and Ground Service*, and attributes that enhance passengers' feelings such as 'Great', 'Excel-lent/good/nice', 'Thank'. The word cloud results also suggest dimensions related to seat comfort and cabin cleanliness such as 'Comfortable' and 'Clean', in which the attribute was not significant in the logistic model.

In conclusion, this study is able to confirm linkage between recommendation scores and text comments. The results suggested airlines should reinforce their trainings to improve human interactions such as personality and friendly services, while negative comments suggested frequent operational improvements such as on-time performance and seat comfort. This study also demonstrates how NPS could be used in instead of binary 'recommended' scores in predicting tendency for recommendations. This study also sheds a light on service operations research on using real-time web scraping technique to monitor performance of the service and to predict recommendation outcome which is rare in service context. This study has elevated the nature of service operations study to prioritize desirable service dimensions that both confirmed by passengers from qualitative and quantitative sources.

References

1. Jiang, H., Zhang, Y.: An investigation of service quality, customer satisfaction and loyalty in China's airline market. J. Air Transp. Manage. **57**, 80–88 (2016)
2. Hapsari, R., Clemes, M.D., Dean, D.: The impact of service quality, customer engagement and selected marketing constructs on airline passenger loyalty. Int. J. Qual. Serv. Sci. **9**, 21–40 (2017)
3. Brochado, A., Rita, P., Oliveira, C., Oliveira, F.: Airline passengers' perceptions of service quality: themes in online reviews. Int. J. Contemp. Hospitality Manage. **68**, 61–75 (2019)
4. Tsafarakis, S., Kokotas, T., Pantouvakis, A.: A multiple criteria approach for airline passenger satisfaction measurement and service quality improvement. J. Air Transp. Manage. **68**, 61–75 (2018)
5. Siering, M., Deokar, A.V., Janze, C.: Disentangling consumer recommendations: explaining and predicting airline recommendations based on online reviews. Decis. Support Syst. **107**, 52–63 (2018)

6. Chatterjee, S.: Explaining customer ratings and recommendations by combining qualitative and quantitative user generated contents. Decis. Support Syst. **119**, 14–22 (2019)
7. Pavela, J., Suresh, R., Blue, R.S., Mathers, C.H., Belalcazar, L.M.: Management of diabetes during air travel: a systematic literature review of current recommendations and their supporting evidence. Endocr. Pract. **24**(2), 205–219 (2018)
8. So, K.K.F., King, C., Hudson, S., Meng, F.: The missing link in building customer brand identification: the role of brand attractiveness. Tour. Manag. **59**, 640–651 (2017)
9. Punel, A., Hassan, L.A.H., Ermagun, A.: Variations in airline passenger expectation of service quality across the globe. Tour. Manag. **75**, 491–508 (2019)
10. Seiler, R., Müller, S., Herzog, C.: Choice overload In the airline industry: a web experiment on booking flights online. In: 47th EMAC Annual Conference, Glasgow, 29. Mai–1. Juni 2018 (2018)
11. Demo, G., Rozzett, K., Fogaça, N., Souza, T.: Development and validation of a customer relationship scale for airline companies. BBR. Braz. Bus. Rev. **15**(2), 105–119 (2018)
12. Reichheld, F.F.: The one number you need to grow. Harvard Bus. Rev. **81**(12), 46–55 (2003)
13. Keiningham, T.L., Aksoy, L., Cooil, B., Andreassen, T.W., Williams, L.: A holistic examination of Net Promoter. J. Database Mark. Customer Strategy Manage. **15**(2), 79–90 (2008). https://doi.org/10.1057/dbm.2008.4
14. Deshpande, M., Sarkar, A.: BI and sentiment analysis. Bus. Intell. J. **15**(2), 41–49 (2010)
15. Kristensen, K., Eskildsen, J.: The validity of the Net Promoter Score as a business performance measure. In: 2011 International Conference on Quality, Reliability, Risk, Maintenance, and Safety Engineering, pp. 970–974. IEEE (2011)
16. Temple, J.G., Burkhart, B.J., Tassone, A.R.: Does the survey method affect the net promoter score? In: Ahram, T., Falcão, C. (eds.) AHFE 2019. AISC, vol. 972, pp. 437–444. Springer, Cham (2020). https://doi.org/10.1007/978-3-030-19135-1_42
17. Seth, S., Scott, D., Svihel, C., Murphy-Shigematsu, S.: Solving the mystery of consistent negative/low net promoter score (NPS) in cross-cultural marketing research. Asia Mark. J. **17**(4), 43–61 (2016)
18. Pollack, B.L., Alexandrov, A.: Nomological validity of the Net Promoter Index question. J. Serv. Mark. **27**, 118–129 (2013)
19. Atalık, Ö., Bakır, M., Akan, Ş.: The role of in-flight service quality on value for money in business class: a logit model on the airline industry. Adm. Sci. **9**(1), 26 (2019)
20. Hu, H.-Y.: Research on customer churn prediction using logistic regression model. In: Xhafa, F., Patnaik, S., Tavana, M. (eds.) IISA 2018. AISC, vol. 885, pp. 344–350. Springer, Cham (2019). https://doi.org/10.1007/978-3-030-02804-6_46
21. Bellizzi, M.G., Eboli, L., Forciniti, C., Mazzulla, G.: Air transport passengers' satisfaction: an ordered logit model. Transp. Res. Procedia **33**, 147–154 (2018)
22. Bhat, V.N.: A multivariate analysis of airline flight delays. Int. J. Qual. Reliab. Manage. **12**, 54–59 (1995)
23. Lapré, M.A., Scudder, G.D.: Performance improvement paths in the US airline industry: linking trade-offs to asset frontiers. Prod. Oper. Manage. **13**(2), 123–134 (2004)
24. Josephat, P., Ismail, A.: A logistic regression model of customer satisfaction of airline. Int. J. Hum. Resour. Stud. **2**(4), 197 (2012)
25. Ahuja, V., Shakeel, M.: Twitter presence of jet airways-deriving customer insights using netnography and wordclouds. Procedia Comput. Sci. **122**, 17–24 (2017)
26. Bird, S., Klein, E., Loper, E.: Natural Language Processing with Python: Analyzing Text with the Natural Language Toolkit. O'Reilly Media Inc., Sebastopol (2009)
27. Nguyen, T.T., Chang, K., Hui, S.C.: Word cloud model for text categorization. In: 2011 IEEE 11th International Conference on Data Mining, pp. 487–496. IEEE (2011)
28. Nohr, E.A., Liew, Z.: How to investigate and adjust for selection bias in cohort studies. Acta Obstet. Gynecol. Scand. **97**(4), 407–416 (2018)

Artificial Intelligence Based Decision Making for Venture Capital Platform

Ankit Tewari[1,2,3], Joaquim Gabarro[1(✉)], Josep Sole[3], Brice Lapouble[3], and Lluis Montull[3]

[1] Universitat Politècnica de Catalunya (UPC), Barcelona, Spain
gabarro@cs.upc.edu
[2] Universidad de Barcelona (UB), Barcelona, Spain
[3] OneRagtime, Carrer de Joaquim Ruyra 9, 08025 Barcelona, Spain
{ankit,josep,lluis}@oneragtime.com

Abstract. Venture Capital as an industry forms the backbone of innovation today. However, despite investing in high technology startups, most of the venture capital firms face a variety of challenges in its day to day operations which lack the adoption of technology. In this work, we introduce a platform aimed at automating some extremely crucial portions of the overall workflow of such firms that ranges from automating the sourcing of opportunities to analyzing and predicting their likelihood of success.

Keywords: Venture capital · Artificial intelligence · Platforms

1 Introduction

Venture Capital firms around the world have been investing in high technology startups and the outcome is the appearance of sophisticated technologies that form the backbone of our lives on a day to day basis [13]. Despite all of this, the very own way of functioning of most of the venture capital firms remains old school leading to lack of availability of interesting opportunities for investment, significantly long time spent in analyzing the opportunities before investment and sometimes human biases in deciding investment worthiness of an opportunities. OneRagtime[1], being a next generation venture capital fund as a *platform* [5] aimes to bring the revolutionary change in the way the venture capital firms work starting from sourcing of opportunities to investment into them.

---

[1] Online; accessed 17 February 2020,https://www.oneragtime.com/about/.

J. Gabarro—Partially supported by MINECO and FEDER grant TIN2017-86727-C2-1-R and AGAUR 2017 SGR 786.
A. Tewari—Supported by OneRagtime S.L.

© Springer Nature Switzerland AG 2020
J. M. Moreno-Jiménez et al. (Eds.): ICDSST 2020, LNBIP 384, pp. 136–149, 2020.
https://doi.org/10.1007/978-3-030-46224-6_11

1.1 Decision Making Process of Venture Capitalists

Venture Capital Investment Decision Making is a fundamental topic [3, 4]. Nowadays, most of the venture capital firms around the world make the decision about investment worthiness based on the following parameters.

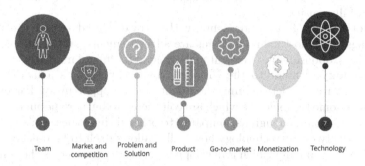

Fig. 1. Company assessment parameters

- **Team:** The quality of team has a strong role to play in deciding how well a company might perform on receiving an investment. In order to determine the quality, we consider a variety of factors such as 'entrepreneurial capability' of founders which helps in deciding if there exists a capability in the founders to successfully lead the team, 'specific talent availability' which means if the company has experts in the team specialized in the area of operation of the company and finally based on these two parameters we determine a team score which decides how well the team is capable together to continue further;
- **Market and Competition:** This factor has an extensive role to play. In the context of analyzing the market and competition, the goal is to determine if there are already significant rival players in the market segment in which the company is going to operate. For example consider the case of 'Ola Cabs', an Indian startup which undergoes our evaluation process. Then, while analyzing it for market and competition, the goal will be to determine how many significant competitors are there in the ride-sharing market (the segment in which it is going to operate in India). By significant, we mean the rivals or competitors which are already present in the same location as this company to offer a stiff competition.
- **Problem and Solution:** Basically, it explores the problem which is going to be addressed by the product developed by the company. For example if a company says that they're going to create a solution for fighting air pollution then the product shall be aimed at helping eliminating the pollutants in the air directly or indirectly.
- **Product:** It determines how well the product is designed considering the solution mentioned above. What is the feasibility of using it, level of complication while using it, adoption of state-of-the-art etc.

- **Go-to-Market-Strategy:** Most often we have seen that some segments of the economy have a dominant go to market strategy. For example consider dating. In this segment, the most common go-to-market-strategy shall be B2C because most often the products in this segment are meant for individual consumers. It is likely that some product may offer B2B2C approach but then the number of users will be significantly less and so will be the chances of successfully raising revenues.
- **Monetization:** It is fundamentally the strategy used to sell the product offerings. The most common forms are advertisements, subscriptions, in-app purchases, sponsorships, referral marketing and affiliate marketing.
- **Technology:** Similar to the Product, the Technology as a factor plays a significant role in determining the quality of the opportunity. For example, there are some technologies which are considered to be more popular as well as relatively mature enough as compared to other technologies. This popularity and maturity of the technology how well it allows itself to be quickly adopted by both the company which is developing the product as well as the consumers who will be the end users.

In this study, we aim at automating some of the most crucial aspects of decision making which includes automating the sourcing of opportunities, performing automated market and competition research and determining the likelihood of success of an opportunity based on one or more of the above mentioned parameters. We develop a Data Science approach [2,6,8]. The rest of the work is organized into three sections. In the Sect. 2, we discuss the major drawbacks in the present methods of decision making and suggest novel methods. It will be followed by Sect. 3 where we will showcase major results explaining the capability of our systems. Finally, we will discuss the conclusions of this study and the scope of conducting future research into evolving the platform.

2   Machine Intelligence for Venture Capital Platforms

2.1   Deal-Sourcing System

The idea of developing a deal sourcing system is based on augmenting the capability of human analysts to be able to source the deals from a wider and more broad pool of sources which would otherwise be requiring a large workforce or would be simply impossible task for small venture capital firms.

Deal sourcing is generally considered a complex and time consuming process. The main reason behind this argument is the extensive need of networking required to generate good deals where the extent of networking ranges from the intern to the managing partner. This kind of networking often involves other venture capital firms and sometimes entrepreneurs also apart from the general source of networking which is the startup ecosystem itself (events, talks, meetups, etc.). In order to keep the deal flow smoothly running and to meet and maintain the relations with Merger and Acquisition boutiques so they keep sending opportunities, it is of utmost importance to focus on reducing the time consumed in finding relevant deals overall.

Therefore, it becomes extremely crucial to see the new deals as quick as possible to invest in the right opportunity before anyone else plans to do. This investment allows the startup to develop the technology faster than the rivals and therefore increases the probability of the high returns for the investor.

In the quest of developing automated systems for deal sourcing, the major challenge is mainly three fold i.e. precisely extraction of data, extraction of relevant entities from the data and finally, the extraction of associated information about that entity. In this sequence, we propose a fully automated deal-sourcing system which consists of the following layers.

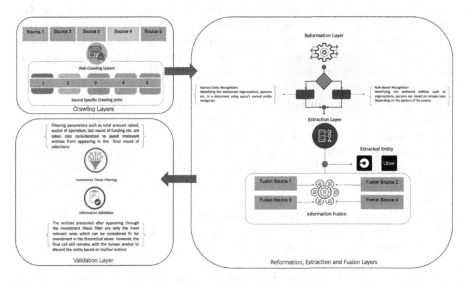

Fig. 2. Deal sourcing system overview

1. Mapping: It is the first phase of the operational deal sourcing system architecture which is used to create mapping between the sources of the data and the framework where rest of the treatment takes place.
2. Crawling: Once the sources are identified and a formal mapping is created between the data origin/source and the data crawling layer which is tasked to extract the data present on the source, source specific crawlers are used to extract the information present on the particular source. However, depending on the method of data presentation on the source, the crawled data can be extracted in any format such as Text, Hypertext Markup Language (HTML), Multipurpose Internet Mail Extensions (MIME) etc. While it is a great advantage to be able to extract data in any of the formats, it also presents with a challenge of non-homogeneity of data which complicates the task of making sense of data based on its format.

3. Reformation: In order to resolve the conflict created by non-homogenous type of data, there are formatters defined to process the incoming data into a standard accepted format. The formatters feed the homogenous form of data through a pipeline so that they can be sent to the next layer which is the extraction layer.

4. Extraction: The role of extraction layer is to detect the entities of interest. In the case of deal sourcing, we are specifically concentrated on detecting the organizations and persons of interest. The functional architecture of the extraction layer is a combination of both artificial intelligence based and rule based systems. Together these systems are used to make sense of the data using spacy's pre-trained named entity recognition algorithms acting as the AI component and well defined rules created using regular expressions acting as the rule-based component. The outcome of the process is a collection of extracted entities mentioned in the data. The functional architecture of the system can be well understood by referring to the Fig. 2.

Clearly, we can observe that the homogeneous data coming from the reformation layer passes through a decision block which decides whether the detection of the entities for that specific source should be artificial intelligence based or rule based and based on the output, the concerned system is called for extracting the entities. In a similar manner, if the decision block suggests the application of rule based component for entity detection, the specifically created rules using the regular expressions are used to extract the entities of interest from the input sentence.

5. Fusion: The output from the previous layer is used to perform data fusion from many other sources in order to improve the data quality and at time provide more visibility about the extracted entity which can be a person or an organization in most of the cases.

6. Investment Thesis Validation: Most of the venture capital firms/organizations have a strictly defined investment thesis. An investment thesis is a framework which maps an abstract idea to a concrete investment strategy. In general, it means that while some venture capital firms invest only in early stage organizations which are going for seed round of funding (say), there might be others which are more liberal while investing and may invest in a Series-A or later rounds of funding comfortably. Now, once enough data is generated about an entity, it becomes extremely essential to determine if that entity fits in the investment thesis of the venture capital firm. It is done to avoid flooding the deal sourcing platform with irrelevant information which instead of simplifying the deal sourcing process, ends up in confusing the analyst.

2.2   Market and Competition

Competitive intelligence [9] is a valuable information that allows us to have a complete picture of competing organizations or individuals in a particular ecosystem. Therefore, there had been earlier approaches to provide competitive intelligence about an organization or individual using a variety of means and the most common solution of all was to exploit crowdsourcing as a means to acquire data. This approach was exploited by Owler [14], a company founded by Jim Fowler and Tim Harsch in 2011[2]. Owler allows a large number of business professionals to enter information about the rivals of companies where they work. It creates a directed a graph of relation between the company of the individual and each of the rivals. However, while being inexpensive and fast to create and retrieve data, such systems suffer from a variety of drawbacks such as-

1. Quality concerns: Many times, it is observed that the information entered manually is prone to errors i.e. either not so accurate or contains human biases. It happens because human perceptions may differ specially if there are many crowdsourcing analysts. For example, it might be possible that one analyst considers Facebook as the biggest rival of Google considering the revenues while another may disagree on it considering the reason that Facebook is eventually not a search engine while Yahoo, DuckDuckGo, Bing etc. are potential rivals.

2. Time concerns: Humans often find about the rival entities by spending enormous time on search engines that lead to certain clues. However, machines are capable to analyze millions of documents in fractions of seconds to understand how close two entities are to become rivals.

Mathematically speaking, the idea of finding the rival companies for a startup is a classic example of a content based recommender systems. Formally, a recommender system is defined as an information filtering system which allows us to tackle the problem of data explosion by filtering and presenting only the relevant information concerned to the user from a pool of data [7,10,11]. In the absence of such system, the user would have to manually scan the data to find the relevant information and depending on the type and size of the data, the problem of finding relevant information manually would have been proportionately difficult. It is exactly what happens when an analyst uses web search with varying queries to gather competitive intelligence.

In order to create a content based recommender system, we exploit the information about all the startups in our database such as their descriptions and the sector in which they operate and create a distance matrix where there are as many rows in the matrix as there are startups in our database and the same applies to the number of columns i.e. we measure distance of startup to every other startup. Further, we use this distance matrix in a nearest neighbour model to find the top K nearest neighbours of every startup under our consideration.

Now, in order to create the distance matrix, we transform textual information about a startup into a numeric vector (which is essential to create a distance matrix) using the term-frequency-inverse-document-frequency (henceforth, tf-idf) method [1] (pag 30) for extracting numeric features out of the text document which is consisting of description and sector of a startup. In order to decide the vocabulary (i.e. features corresponding to each data point) of the tf-idf vectorizer, we are using all the sectors in our database to create a tokenized vocabulary. Eventually, we have a dataframe where each row describes a particular startup

---

[2] https://corp.owler.com/, Online; accessed 18 February 2020.

accompanied by a number of features corresponding to the sectors represented as features.

Before we can dig deeper into the outcome of our recommendation system, let us understand the tf-idf based method for feature extraction. The tf-idf is a score computed using two different metrics i.e. term-frequency (tf) and inverse-document-frequency (idf). In order to define the specific meaning of each of these documents, let us assume that we have N documents present in our collection (here, each document attempts to merely describe a particular startup using three features (i.e. short_description, description and sectors). Now, let us consider that the frequency of occurrence of word i in document j is f_{ij}. Then, we can say that the TF_{ij} is defined as $TF_{ij} = f_{ij}$.

However, we normalize this score by dividing it by the frequency of the most frequent i.e. most common word in the document (by taking care that it is not a stopword i.e. words such as such as "the", "a", "an", "in"). Therefore, the most frequently occurring term is assigned the *term frequency* of 1 while all other terms have scores relative to this term, $TF_{ij} = f_{ij}/\max_k f_{kj}$.

Similarly, we compute the *inverse document frequency* of a term by considering the frequency of occurrence of a term across different documents in our collection i.e. suppose the term i occurs in exactly n_i documents out of the total N documents in our collection then the *inverse document frequency* score of the term can be defined as $IDF_i = \log_2(n_i/N)$.

Therefore, the *term frequency-inverse document frequency* score of a term may be defined as $TF.IDF_{ij} = TF_{ij} \cdot IDF_i$. It allows us to conclude that higher the tf-idf score of a sector correspoding to a startup, the more significant role it plays in describing the content of the document.

Now, once we have all the data in a dataframe where each startup has vectorized representation, we can begin modelling using the *K-Nearest Neighbors* (KNN) algorithm [2,6]. This algorithm is about finding K most similar neighbors for a given item. In the context of recommending startups, each of the startup is treated as an individual row and the features describing it are the sectors we used as vocabulary. The problem of KNN can take different forms example, KNN search or KNN graph construction. While KNN search aims to find the K nearest neighbors of only a small number of items in the collection of say, N items, the KNN graph construction does it for all the items in the collection. Formally, a K-nearest neighbor (kNN) graph of a system is a directed graph where each node is connected to its K most similar neighbors where distance is measured by a predefined metric. Some of these similarity metrics are the *manhattan distance*, $\Sigma\|x - y\|_{i=1}^{N}$ or the *euclidean distance*, $\sqrt{\Sigma(x - y)^2}_{i=1}^{N}$ where the first point is $(x_1, x_2, x_3, ..., x_n)$ and the second point is $(y_1, y_2, y_3, ..., y_n)$.

Finally, we evaluate the nearest neighbours of a startup obtained both by manual identification of significance as well as by measuring similarity using the metric $Cosine\_Similarity = |U.V|/(\|U\|\|V\|)$.

2.3   Opportunity Evaluation System

We can clearly observe that the factors such as team quality, product, technology, go-to-market-strategy, monetization strategy and market and competition play a key role in determining investment worthiness. We have used the team count, members count and fundraising count as fundamental features to estimate the success of a startup. Although, the success of a startup is a highly subjective topic but in the case of our investment thesis, we have assumed that any startup that has raised funds beyond series B shall be considered a success. Further, in addition to these 3 features, we have used tf-idf to create 50 new features for each of the startups.

Eventually, we have decided to fit different models to begin experimenting with the performance of those models on the dataset consisting of 53 features. In Sect. 3, we discuss some results obtained using the *Random Forest Classifier* and *Gradient Boosting Classifier* [2,6] with default parameters to decide our candidate model. The main reasons for choosing ensemble based models were the inability of Logistic Regression model's solvers to optimize the model fitting process and relatively high dimensionality of the dataset which probably posed problems for the linear model.

3   Experiments and Results

3.1   Deal-Sourcing System

The Fig. 3 describes the outcome of the deal sourcing system without examining the outcomes of individual layers of the system. Consider that we have crawled a source and identified the following sentence through the initial layers of system-

SnapLogic Secures \$72 Million in Growth Financing Led by Arrowroot Capital. Norwest Venture Partners participates in Dave's \$50M funding round.

In the advance layers, it is served as an input to the spacy's pre-trained model" $en\_core\_web\_lg(v2.2.0)$ " and the outcome is a tagged version of the input where the tags correspond to the entities of interest (which might be organizations or persons mainly in our case) mentioned in these sentences. In order to examine the system in terms of the runtime of individual layers, we performed two runs of each of the layers to extract the information from the same input sentence as mentioned above. The results are mentioned in Table 1.

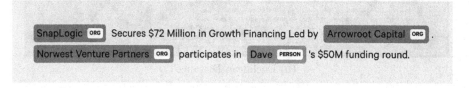

Fig. 3. Results of Named Entity Recognition

Table 1. Summary of runtimes of different modules during API calls of deal sourcing system (milliseconds in a laptop)

Stage	Attempt I	Attempt II
Crawling	1234	1435
Reformation	762	813
Extraction	704	681
Fusion	3403	2844
Validation	1838	1788

3.2 Market and Competition

The Fig. 4 explains the performance of the market and competition research system. Figures 4a and b are having the recommended startups (in green) that are similar to the ride-sharing startup 'Uber'. In order to facilitate human inspection, the sectors in which they operate are mentioned in front of their names (in white). The nearest neighbour model used to recommend these startups was allowed to run with two different distance metrics i.e. Manhattan distance (Fig. 4a) and Euclidean distance (Fig. 4b). We can clearly observe by visually inspecting the sectors of recommended startups that the performance of Euclidean distance is better than Manhattan distance as the former have all the 11 recommended startups falling in the category of ride-sharing platforms while the latter has only

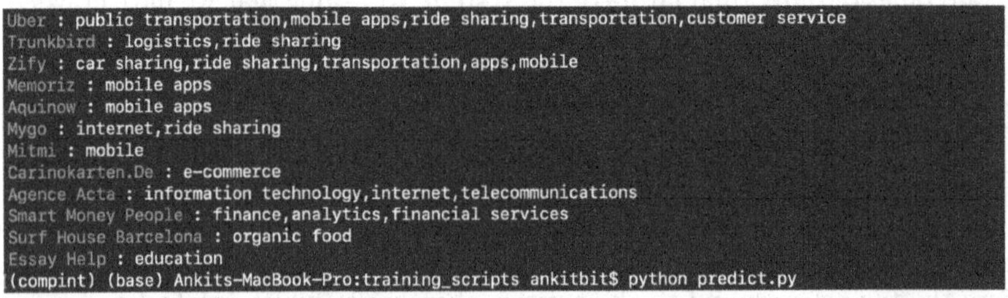

```
Uber : public transportation,mobile apps,ride sharing,transportation,customer service
Trunkbird : logistics,ride sharing
Zify : car sharing,ride sharing,transportation,apps,mobile
Memoriz : mobile apps
Aquinow : mobile apps
Mygo : internet,ride sharing
Mitmi : mobile
Carinokarten.De : e-commerce
Agence Acta : information technology,internet,telecommunications
Smart Money People : finance,analytics,financial services
Surf House Barcelona : organic food
Essay Help : education
(compint) (base) Ankits-MacBook-Pro:training_scripts ankitbit$ python predict.py
```

(a) Manhattan Distance

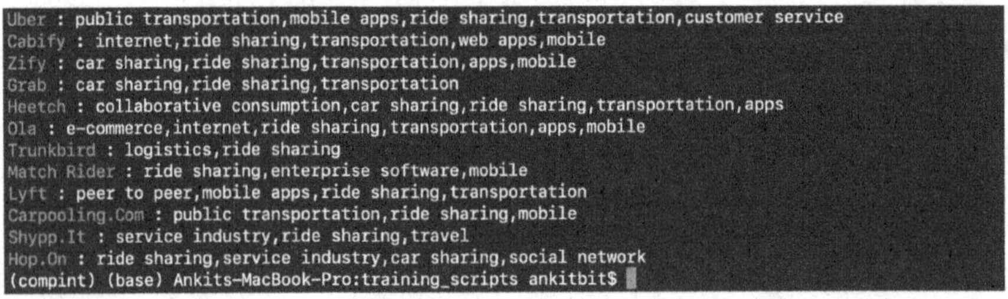

```
Uber : public transportation,mobile apps,ride sharing,transportation,customer service
Cabify : internet,ride sharing,transportation,web apps,mobile
Zify : car sharing,ride sharing,transportation,apps,mobile
Grab : car sharing,ride sharing,transportation
Heetch : collaborative consumption,car sharing,ride sharing,transportation,apps
Ola : e-commerce,internet,ride sharing,transportation,apps,mobile
Trunkbird : logistics,ride sharing
Match Rider : ride sharing,enterprise software,mobile
Lyft : peer to peer,mobile apps,ride sharing,transportation
Carpooling.Com : public transportation,ride sharing,mobile
Shypp.It : service industry,ride sharing,travel
Hop.On : ride sharing,service industry,car sharing,social network
(compint) (base) Ankits-MacBook-Pro:training_scripts ankitbit$
```

(b) Euclidean Distance

Fig. 4. Recommendations with two distances

3 relevant recommendations. Further, we have also used the cosine similarity to explain the average of cosine similarities between Uber's attributes such as description and sectors and each of the its recommended competitors. If we compute the cosine similarity between the Uber and each of the top 6 competitors, then it comes out that in the Table 2 that Euclidean distance outperforms the manhattan distance using both measurement parameters i.e. while comparing recommendations only based on description and while comparing recommendations based on both description and sectors.

Table 2. Summary of distance metrics and mean cosine similarity scores

Distance metric	Only description	Description and sectors
Manhattan distance	0.4780	0.6585
Euclidean distance	0.5075	0.7120

3.3 Opportunity Evaluation

We can observe the relative better performance of Random Forest model from the Figs. 5a and b which are describing the classification reports of the Random Forest and Gradient Boosting models respectively. Since both of these models were allowed to run with default parameters, it creates a level field for playing for both the algorithms. While the Random Forest based model attained a f1-score of 0.86 and 1.00 for the classes closed and survived, the Gradient Boosting model was able to attain the f1-score of 0.75 and 1.00 for the same classes respectively.

```
                precision    recall  f1-score   support

      closed       0.77      0.99      0.86       290
    survived       1.00      1.00      1.00     41250

    accuracy                           1.00     41540
   macro avg       0.88      0.99      0.93     41540
weighted avg       1.00      1.00      1.00     41540
```

(a) Random Forest

```
                precision    recall  f1-score   support

      closed       0.64      0.90      0.75       199
    survived       1.00      1.00      1.00     29623

    accuracy                           1.00     29822
   macro avg       0.82      0.95      0.87     29822
weighted avg       1.00      1.00      1.00     29822
```

(b) Gradient Boosting

Fig. 5. Classification reports

Further, we can observe the relative better performance of the Random Forest model over the Gradient Boosting model by considering the Fig. 6 of confusion matrices for the two models. We can clearly observe that while both the models perform similar for the true positive errors, the Random Forest classifies the true negatives in 71% of the cases while Gradient Boosting was able to do so only for 48% of the cases. Therefore, following the aforementioned facts, we have decided to continue with Random Forest in order to tune the parameters and examine the performance of the model on the test set. Once parameter tuning was performed on the Random Forest model, we observed that there was no difference between the base model and the tuned model except for the boot-strap = False, max_depth = 80, min_samples_split = 10 and n_estimators = 1000 is the optimized model as compared to bootstrap=True, max_depth=None, min_samples_split=2, n_estimators=10 in the base model. Further both the models achieve a training accuracy of 99.9% (if we take one significant digit after decimal notation) which makes no real difference between base model and the optimized model.

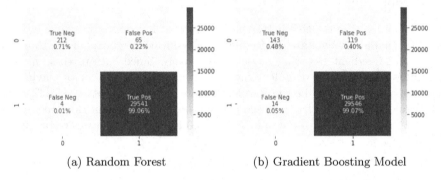

(a) Random Forest   (b) Gradient Boosting Model

Fig. 6. Confusion matrices

In order to examine the validity of the model, we decided to investigate it in terms of its feature significance values. Feature significance is a global level interpretability method. Globally, we can interpret the model in terms of the values of global parameters which were learned while fitting the model and feature importance is one such metric that can be used to explain on the global level which features were the most significant ones that facilitated the model in learning the patterns accurately. The Fig. 7 describes the significance of features in determining the decisions at the global level. We can observe clearly that members_count is the most important feature which is helping the model in learning patterns accurately. It can be interpreted from the usual understanding that an organization having employed a large number of members is more likely to survive as compared to one having fewer members based on the fact that it must be having adequate financial resources to support those members in terms of salaries, perks etc.

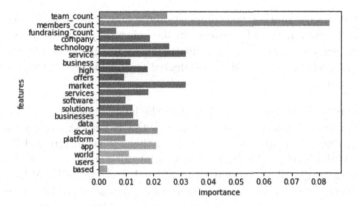

Fig. 7. Feature importance: global explanation of random forest

4   Conclusions

In this work, we have identified three different areas concerned with the application of artificial intelligence in the domain of venture capital. A extended version can be found in [12]. These three areas were machine based deal-sourcing for venture capital firms, market and competition research for sourced deals for examining their relevance and for opportunity evaluation for the deals which are converted into opportunities i.e. determining the potential of a startup to raise funds beyond Series B round of funding.

In order to understand better, similar to a sales cycle, we have created our investment cycle where the deal-sourcing system is assigned the responsibility to source deals which are difficult to source manually. In the context of investment, these deals are just probable startups which are being incubated or accelerated at an incubator or accelerator or in some cases just born. Now, once the deals are identified, we use our competitive analysis system which filters the deals based on the market and competition factors. The deals which survive through this stage become investment prospects and they are finally analyzed through our opportunity evaluation system which determines the investment worthiness of the prospects and based on a positive outcome, these prospects become opportunities.

In the context of deal-sourcing, we discussed how we may exploit the capabilities of deep learning based natural language processing tools such as Named Entity Recognition to make the best use of technology in sourcing deals which may be potential investment opportunities. While it allows us to analyze any chunk of text obtained from various sources to detect possible entities of interest, it sometimes misclassify an entity of one type (say, organization) into an entity of other type (say, person). Such systems therefore could be improved by training on our own corpora instead of using a pre-trained model. There is enormous scope to employ deep learning in assisting such similar tasks involving analyzing of large number of documents which would otherwise be difficult manually to perform.

Similarly, for the market and competition research system, we have employed the nearest neighbour model. However, while being simple it always comes with may challenges which must be met in nearby future to achieve better than human accuracy and reliability. We see some degree of improvement of our models in the following terms.

1. **Interpretability:** The models based on the idea of nearest neighbours suffer from the lack of interpretability. Since in the K-nearest neighbours model, there are no parameters to be learnt, so there is no interpretability at the very modular level [8]. Further, a model's interpretability can be discussed in the context of global interpretability and local interpretability. While there no parameters to be trained, there are no global parameters to be optimized and so the model behaves locally. Further if we consider interpreting the model locally then we have to explain how an individual instance is having a set of other instances as its neighbours. In this regard, while we can extract the features representing that instance but if their dimensionality is significantly high, it becomes almost impossible for humans to understand them. Perhaps there could be some scope of interpreting the model locally if we take into account some dimension reduction method.

2. **Dimensionality:** Curse of Dimensionality: If the dimensionality of the data is extremely high, the model becomes computationally expensive to find nearest neighbours because it is an instance/memory based algorithm.

3. **Sensitivity:** Nearest neighbours are always scale sensitive which means that if we add additional features in the data which are not normalized then it impacts significantly the system's capability to find good neighbours.

4. **Skewness:** Nearest neighbours based implementation is not effective on rare event (skewed) variable. Therefore, if there is a feature in tf-idf which occurred only for 1–2 particular startups then it will be extremely difficult to find right neighbours for them instead if they have generic terms, a lot of garbage recommendations can be produced which has to be checked.

We can also interpret a model locally based on how a particular instance is classified. That is, which are the most important features that helped in classifying a particular instance into one of those two categories. In this sequence we can use algorithms such as LIME (Local Interpretable Model-Agnostic Explanations) to arrive at a local explanation of each of the attributes which are also model agnostic.

Acknowledgement. Ankit Tewari is extremely grateful to Mr. Om Prakash Tewari for his continuous guidance and always motivating words towards creating systems for the social good.

References

1. Baeza-Yates, R., Ribeiro-Neto, B.: Modern Information Retrieval. Addison-Wesley/ACM Press, Boston (1999)

2. Burkov, A.: The Hundred-page Machine Learning Book. Dover Books on Mathematics, Amzon, Poland (2019)
3. Fried, V.H., Hisrich, R.D.: Toward a model of venture capital investment decision making. Financ. Manag. **23**(3), 28–37 (1994). https://doi.org/10.2307/3665619
4. Harroch, R.: A guide to venture capital financings for startups. https://www.forbes.com/sites/allbusiness/2018/03/29/a-guide-to-venture-capital-financings-for-startups/#6f3f33d51c9c. Accessed 19 Feb 2020
5. Hartnett, H.: The rise of "The Platform" for Venture Capital funds. https://www.forbes.com/sites/heatherhartnett/2017/09/28/the-rise-of-the-platform-for-venture-capital-funds/#26f9a5184484. Accessed 18 Feb 2020
6. Hastie, T., Tibshirani, R., Friedman, J.: The Elements of Statistical Learning - Data Mining, Inference, and Prediction. Springer, New York (2009). https://doi.org/10.1007/978-0-387-84858-7
7. Konstan, J.A., Riedl, J.: Recommender systems: from algorithms to user experience. User Model. User-Adap. Inter. **22**(1), 101–123 (2012). https://doi.org/10.1007/s11257-011-9112-x
8. Molnar, C.: Interpretable Machine Learning - A Guide for Making Black Box Models Explainable. Lulu (2019). https://christophm.github.io/interpretable-ml-book/. Accessed 18 Feb 2020
9. Porter, M.E.: Competitive Strategy: Techniques for Analyzing Industries and Competitors. Free Press, New York (1980)
10. Resnick, P., Varian, H.R.: Recommender systems. CACM **40**(3), 56–58 (1997). https://doi.org/10.1145/245108.245121
11. Ricci, F., Rokach, L., Shapira, B.: Introduction to recommender systems handbook. In: Ricci, F., Rokach, L., Shapira, B., Kantor, P.B. (eds.) Recommender Systems Handbook, pp. 1–35. Springer, Boston, MA (2011). https://doi.org/10.1007/978-0-387-85820-3_1
12. Tewari, A.: Disrupting Venture Capital Through Artificial Intelligence. Master's thesis, Facultat de Matemàtiques i Estadística of the Universitat Politècnica de Barcelona, Master in Statistics and Operations Research UPC-UB, January 2020
13. Wertz, J.: How startup investors can utilize AI to make smarter investments. https://www.forbes.com/sites/jiawertz/2019/01/18/startup-investors-utilize-ai-smarter-investments/#6ea030564ef5. Accessed 18 Feb 2020
14. WikiMili: Owler. https://wikimili.com/en/Owler. Accessed 18 Feb 2020

Supporting Operational Decisions on Desalination Plants from Process Modelling and Simulation to Monitoring and Automated Control with Machine Learning

Fatima Dargam[1]([⊠]), Erhard Perz[1], Stefan Bergmann[1], Ekaterina Rodionova[1], Pedro Sousa[2], Francisco Alexandre A. Souza[2], Tiago Matias[2], Juan Manuel Ortiz[3], Abraham Esteve-Nuñez[3], Pau Rodenas[3], and Patricia Zamora Bonachela[4]

[1] SimTech Simulation Technology, Graz, Austria
{f.dargam,e.perz,s.bergmann,e.rodionova}@simtechnology.com
[2] OnControl Technologies, Coimbra, Portugal
{pedro.sousa,francisco.alexandre,
tiago.matias}@oncontrol-tech.com
[3] IMDEA Water, Alcalá de Henares - Madrid, Spain
{juanma.ortiz,abraham.esteve,pau.rodenas}@imdea.org
[4] Aqualia – FCC Group, Madrid, Spain
patricia.zamora@fcc.es

Abstract. This paper summarizes some of the work carried out within the Horizon 2020 project MIDES (MIcrobial DESalination for low energy drinking water) (The MIDES project (http://midesh2020.eu/) has received funding from the European Union's Horizon 2020 research and innovation programme under grant agreement N° 685793 [1].), which is developing the world's largest demonstration of a low-energy system to produce safe drinking water. The work in focus concerns the support for operational decisions on desalination plants, specifically applied to a microbial-powered approach for water treatment and desalination, starting from the stages of process modelling, process simulation, optimization and lab-validation, through the stages of plant monitoring and automated control. The work is based on the application of the environment IPSEpro for the stage of process modelling and simulation; and on the system DataBridge for automated control, which employs techniques of Machine Learning.

Keywords: Operational decision support · Desalination plants · Process modelling · Process simulation · IPSEpro · Plant monitoring · Automated control · Machine Learning · Horizon2020 project · MIDES · Microbial Desalination Cell · MDC · Low-energy process · Treated wastewater · Drinking water · Climate change adaptation · Sustainability

1 Introduction

According to [2] and [3], the water demand situation, with increasing population and economy growth is estimated to reach 6,900 billion m^3 by 2030. Figure 1 illustrates the estimation of water demand for 2030, considering the demands for agriculture,

© Springer Nature Switzerland AG 2020
J. M. Moreno-Jiménez et al. (Eds.): ICDSST 2020, LNBIP 384, pp. 150–164, 2020.
https://doi.org/10.1007/978-3-030-46224-6_12

industry, domestic sectors
- taking into account the
demand for potable water
increases annually by 2%
[2]. This study shows that if
business-as-usual practices
are continued, it will result
in a global water demand that
is 40% higher than the available water supply. Such estimation establishes well the
importance of studying solutions based on desalination
technologies, which com-
poses the scope of the Horizon 2020 project MIDES

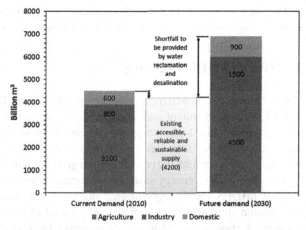

Fig. 1. Global water demand gap between 2010 and 2030 [2].

(MIcrobial DESalination for low energy drinking water) [1], upon which the presented
developments of the current paper are based.

The MIDES project (running from 2016 to 2020) aims to revolutionize energy-
intensive Reverse Osmosis (RO) desalination systems by demonstrating sustainable
production of fresh water at three pilot locations in and outside Europe. MIDES is
developing the world's largest demonstration of a low-energy system to produce safe
drinking water via the use of Microbial Desalination Cells (MDC), which remove ions
from saltwater in an innovative process powered by electroactive bacteria. In a general
basis, the project supports: (i) Climate change mitigation by lowering greenhouse gases
from current desalination systems; as well as (ii) Climate change adaptation through
innovation in providing freshwater resources.

This paper describes the part of the work carried out within MIDES project, concerning the support for operational decisions on desalination plants, specifically applied to
a microbial-powered approach for water treatment and desalination (further outlined in
Sect. 2). The supported operational decisions and actions on the implementation of the
desalination technology, follow a roadmap that starts from the stages of process modelling, process simulation, optimization and lab-validation; through the stages of plant
monitoring and automated control. For the stage of process modelling and simulation,
the developed work is based on the use of the environment IPSEpro [11–14], while for
the stage of automated control, the system DataBridge is applied, employing techniques
of Machine Learning [10].

The current paper is further structured in the following way: Sect. 2 presents the
overall MIDES project process; Sect. 3 describes the MIDES roadmap process from
Lab to Pilot stages. Section 4 presents the overall process modeling and simulation
performed for the MIDES technology using IPSEpro, including the development and
implementation of the cloud-based platform that deployed and shared the MIDES process
online. Section 5 presents the monitoring and automated control stages of the project
(including the Machine learning approach). Section 6 discusses the operational decision

making and decision support in the implementation of the MIDES Pilot Desalination Plant. Finally, Sect. 7 draws some conclusions about the work carried out.

2   Overall MIDES Process

The EU Horizon2020 MIDES project [1] specifically applies a microbial-powered approach for water treatment and seawater desalination. As in [4], existing desalination technologies require high-energy input, being Reverse Osmosis (RO) the most widely used technology for seawater desalination with an energy consumption of at least 3 kWh/m^3. In this context, the project aims to go beyond the state of the art in desalination by developing a low-energy sustainable process: Microbial Desalination Cell (MDC).

The MIDES low-energy-powered approach combines MDC technology as pre-desalination step in connection with conventional RO aiming at increasing desalinated water production, while maintaining low energy requirements. The MIDES overall process (as illustrated in Fig. 2) includes a pre-treatment of the saline stream by ceramic membranes prior to entering the MDC unit, where it is partially desalinated (70-90%) before the RO post-treatment.

As stated in [4], the initial treatment of municipal wastewater in a conventional anaerobic reactor produces an acetate-rich effluent as a fuel for the MDC. In the case of saline water, in conventional RO desalination, seawater or brackish water undergoes several pretreatment steps[1] to protect the membranes from pollutants. In MIDES, ceramic submerged membranes substitute this whole pre-treatment, leading to a reduction in chemicals usage and footprint, as well as lowering by 80% the energy demand. The

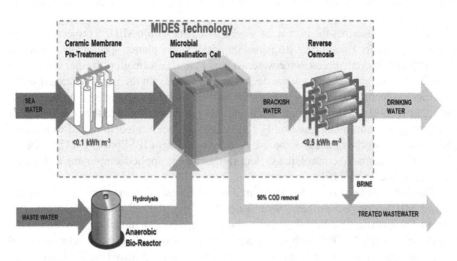

Fig. 2. The overall MIDES process using Microbial Desalination Cell (MDC).

---

[1] For instance: Chemical coagulation, Sedimentation and Filtration.

results obtained with a lab-scale MDC has led to a significant improvement of water production compared to values reported in literature [4].

The overall MDC aspects and its performance in MIDES [1] have been explained in various publications of the project consortium (among them [4, 5, 21]), as well as in most of the project's deliverables (among them [6–10]). The focus of this paper is to report about the developments of the project work-package concerning *Process Simulation & Analysis, Automation & Control* and its implications to the operational decision making process of the overall project implementation and pilot installations.

3   MDC & MIDES Process from Lab to Pilot Scale

As planned in the project, there were three implementation phases in order to test and validate the MIDES MDC and its overall concept. Those three phases included: (1) the Lab-scale MDC; (2) the Pre-Pilot-scale MDC; and (3) the Pilot-scale MDC, including the overall MIDES process in operation.

The roadmap of the implementation of the MIDES technology goes through the testing of the Lab-scale-MDC, up-scaling to the assembly of a Pre-Pilot-MDC, towards the development of the world's largest demonstrator of the innovative MDC technology at Pilot-scale, which will be validated in three demo-sites: Dénia (Spain)[2]; Tenerife – Canary Islands (Spain)[3]; and in a demo-site outside Europe (in Egypt)[4]. All three Pilot plants will be constructed and operated under real environments in desalination plants operated by the MIDES project coordinator Aqualia [4].

3.1   The Lab-Scale MDC

The Lab-scale-MDC was composed of a 100 cm^2 electrode area, with the configuration of the MDC press filter reactor assembling elements, including: plates, gaskets, inlets, outlets, electrodes, and membranes.

Besides all the necessary lab tests and procedures for implementation and installation at Lab-scale stage, the MIDES process components and the overall process model were mathematically modelled, simulated and validated in the environment IPSEpro[5] [12], by implementing the main engineering/physical laws governing the main processes involved in the system - combined physical, chemical and electrochemical principles, having differential equations used due to their strong ability to embody the dynamics of MDC system [6], as well as parameter values and initial conditions setting specific situations to operate the model. A dedicated Desalination/Wastewater-Treatment Model Library with MIDES component-models was created in IPSEpro-MDK [13] to serve this purpose. Figure 3 presents the MDC flow chart used for the mathematical model development, which also illustrates the scheme that was assembled in the Lab.

---

[2] Pilot demo-site in Dénia (Spain) was already launched in November 2019.

[3] Pilot demo-site in Tenerife (Spain) is still under construction and is planned to be launched early in 2020.

[4] The installation schedule of the Pilot demo-site outside Europe still needs to be confirmed.

[5] IPSEpro: http://www.simtechnology.com/CMS/index.php/ipsepro.

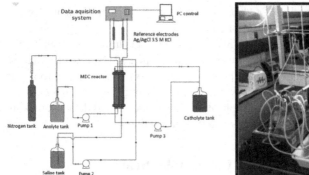

Fig. 3. The MIDES lab-scale Microbial Desalination Cell (MDC).

3.2 The Pre-Pilot MDC

The Pre-Pilot-MDC was assembled comprising 15 modular stack units of

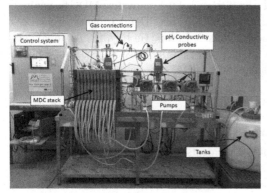

650 cm^2 electrode area per cell. At the pre-pilot stage, apart from all the necessary implementation and installation procedures, the MIDES overall process model was simulated, optimized and validated using IPSEpro [12]; and its automated control stage was built applying the system DataBridge [10].

Figure 4 shows all the components of the complete MDC Pre-Pilot set-up, including its control module. For the control stage linked to the complete MDC set-up of the MIDES pre-pilot system, the

Fig. 4. The MIDES pre-pilot system set-up at IMDEA lab.

parameters monitored were automatically controlled via the system DataBridge from project partner OnControl (as described in Sect. 5).

3.3 The Pilot-Scale MDC

The Pilot-scale MDC specification comprises of an assembly of MDC pilot-unit stacks of 12 pilot MDC unit cells (4 units per pilot plant) of 0.8 m^2 electrode area per cell, 50 × 80 cm. This development represents the world's largest demonstrator of the innovative MDC technology at Pilot-scale. Figure 5 shows the MIDES Pilot-scale MDC at the installations of the Aqualia Desalination Plant in Dénia, Spain. This demo-site was launched in November 2019 and is currently running at operational testing and validation phases, including its controlled operation.

Fig. 5. The MIDES pilot Microbial Desalination Cell (MDC) in Dénia Demo-site.

4   MIDES Process Modelling and Simulation

The tasks of process modelling, simulation, and process optimization were developed for MIDES using SimTech's simulation environment IPSEpro [12, 13].

4.1   IPSEpro – SimTech's Integrated Process Simulation Environment

IPSEpro (SimTech's Integrated Process Simulation Environment) is a heat balance and process simulation software package, which is currently one of the most comprehensive and versatile process modeling systems available. IPSEpro can be applied in a wide range of applications, within the areas of Desalination; Geothermal Energy; Refrigeration; Concentrating Solar Power; and Thermal Power. With its various modules, IPSEpro supports users throughout the entire lifecycle of a process plant, from conceptual design to on-line plant performance monitoring and optimization. Due to its unique level of flexibility and its open architecture, IPSEpro is the ideal platform for implementing custom modelling solutions [12–14] (Fig. 6).

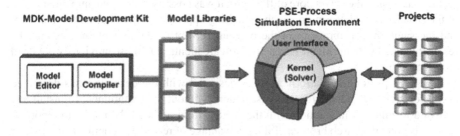

Fig. 6. IPSEpro general architecture.

In IPSEpro, process models are created using the Process Simulation Environment (PSE). In PSE, the user sets up the process model graphically by drawing a flowsheet using components from a model library. Required data is entered directly in the flowsheet. By drawing the flowsheet and entering the data, the user implicitly creates a system of algebraic equations, which is then solved by the IPSEpro's solver core. Results are displayed graphically in the flowsheet [11]. In IPSEpro component equations and physical property methods are not part of the core software. Instead, application-specific information is contained in model libraries, which can be created and modified using a special Model Development Kit (MDK). For instance, a model library specifically developed for ORC (Organic Rankine Cycle) processes [18] includes a comprehensive set of component models based on working fluids used, as well as models for the part-load behaviour of the components, so that the user can analyze the off-design characteristics of ORC plants [16, 17]. The same applies for Desalination plants, using the Desalination Model Library [14].

4.2   IPSEpro in MIDES

IPSEpro Process Simulation Environment (PSE) [12] and IPSEpro Model Development Kit (MDK) [13] were both used in the development of MIDES component models, with the creation of a customized model library called MIDES_Lib, as well as in the creation of the overall process model including the MDC.

The IPSEpro simulated models support understanding the performance of the involved processes. The MDC mathematical simulation model was a useful tool to predict the behaviour of MDC systems in different conditions. Hence, the mathematical process model was able to simulate accurately the quantitative influence of different parameters on the MDC performance. The complete MIDES process was implemented, simulated and optimized in the IPSEpro-PSE Process Simulation Environment [12], including the MDC and the other customized desalination and wastewater-treatment components created using the IPSEpro-MDK [13] for the MIDES_Lib.

4.3   IPSEpro Online Platform - Implementation

A requirement of the project MIDES was to present its overall process model in a shared online platform, in order to allow better collaboration among developers and project partners. By using the solver core of IPSEpro, it was possible to base the implementation of the cloud-based platform on a web modelling approach. Using the same solver core and the same model library as in the original IPSEpro environment ensures that the results obtained with the cloud-based platform are identical to those previously obtained with IPSEpro.

The IPSEpro online platform [9] was designed with the characteristics described above for creating and solving process models, without the requirement to install any software locally. All interaction with the model, from defining the model to reporting results, is done via a web browser. Taking advantage of recent developments in browser technology, in particular HTML5, a browser-based user interface has been developed [19], as included in the general IPSEpro online platform concept in Fig. 7; and in the system architecture at implementation-level, shown in Fig. 8.

The IPSEpro cloud-based simulation platform has been implemented as a web application and as such typically defined as a client-server computer program, which the client (including the user interface and client-side logic) runs in a web browser. Consequently, the components of web applications can be divided into the categories: client-side components and server-side components, and a middleware which manages the communication between client and server. Figure 8 illustrates the overall implementation-level architecture of the IPSEpro cloud-based simulation platform, partially developed under the MIDES project.

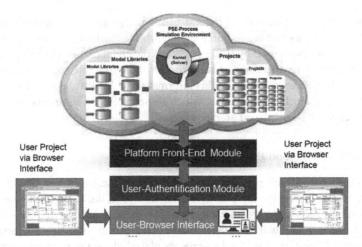

Fig. 7. IPSEpro online platform concept.

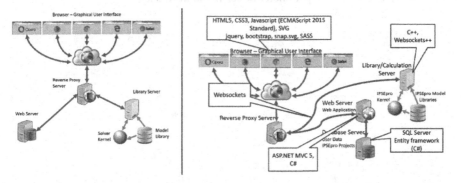

Fig. 8. IPSEpro online platform - system architecture at implementation-level.

Client-Side Component: In the cloud-based simulation platform, the client side comprises of a project management component and the flowsheet editor. The project management component enables the user to log into the system, to create new projects, or select existing projects to use in the flowsheet editor.

Flowsheet Editor: The flowsheet editor enables the user to create and modify the flowsheets of processes, to edit process parameters, to execute the system solver and to display results.

Server-Side Components: The server-side of the cloud-based simulation platform includes: A web server with a database server, which stores all data about users and user projects. The web server also provides all parts of the interface which are not specific to a particular model library; and a Library Server.

Library Server: The Library Server provides two services: (1) upon the respective request from a client, it returns the information about all component in a model library which is required to configure the user interface for using it with the respective model library; and (2) it also handles the request to solve a project. The information about the system components that are required by the flowsheet editor are extracted from the model library description used by IPSEpro. When the client requests to solve a process, it includes with the request a complete description of the structure and parameterization of the system. The library server feeds this information to the solver core, which converts it into a system of algebraic equations, which it then solves. The results are then sent back to the client to be displayed graphically.

4.4   MIDES Desalination Process Model Used Online

Figure 9 shows the MIDES process model within the IPSEpro Online Platform [9]. It is possible to identify on the left side of the environment, the icons of the customized desalination and wastewater-treatment components created for the MIDES_Lib model library and used in the simulation of the overall MIDES process. Results of the simulation are automatically displayed in the data-crosses placed along the process components' connections; and within the parameter boxes in the flowsheet.

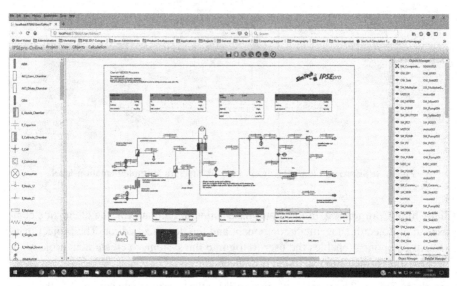

Fig. 9. MIDES process model simulated in the IPSEpro online platform

5   MIDES Monitoring and Automated Control

The MIDES control and monitoring systems were implemented and optimized by the partner OnControl, starting from the field devices up to the high-level control tools. All sensors and actuators were selected considering the principal process variables to be monitored and controlled. Such selection was made based on commercial solutions already available in the market, taking into account the financial viability of the project and future scaling up. Moreover, the actuators and the desired behavior for the process, the low-level control architecture were developed, comprising of a Programmable Logic Controller (PLC) and a Human Machine Interface (HMI). The final element of the control architecture is the high-level control layer that is composed by OnControl software systems used to read, process and visualize data in real time (Fig. 10).

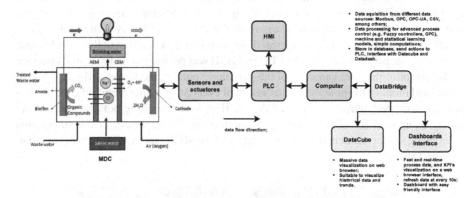

Fig. 10. Control system archicteture.

As illustrated in Fig. 10, the data module Databridge is responsible for all acquisition and processing of the data, which includes all the tools that will provide and expert control of the system. The main Databridge plugins used were the following: (1) Expert Fuzzy Control system: software that allows the incorporation of expert human knowledge about the process in the system to be controlled using "if-then" fuzzy rules. The user can construct the controller and adapt it using a user-friendly graphical interface; (2) Statistical Modeling System; (3) Ontrend Plugin: software that allows the creation of historical data of the process, saving it in a PostgreSQL database; and (4) PostgreSQL: Database used by the Ontrend plugin of the Datacube to show/monitor the historical data of the process.

In addition to Databridge, the systems Datacube and Dashboard were also used in order to achieve unattended control architecture and process monitoring. The Dashboard technology is used for real time process monitoring in a single webpage interface, where the main information from process is made available.

Datacube is a high-level user interface platform that is used for: Historical analysis of the process using the plugin Ontrend; as well as for Analysis of the Key Performance Indicators of the process. The Datacube system is a modular and extensible software, which creates an interface between the operator and the process. It allows for a historical view of the process, online and historical analysis of the energy consumption and automatic creation of reports.

In order to monitor the performance of the MIDES process, a tool for the implementation and monitorization of the Key Performance Indicators (KPI) of the process was created, which handled the most varied KPIs: overall process efficiency, production flow, energy management system, among others. The software offers a graphical interface using the Datacube that allows the user to monitor the KPIs and check its historical data. Such tool is particularly important to support decision-making processes within the project implementation procedures.

5.1 Monitoring Control System in the Pre-pilot

The monitoring automated control system deployed in the MIDES pre-pilot installation within the IMDEA lab set-up is shown in Fig. 11. The development of the control system was an iterative process that starts from the selection of the system requirements, together with the design of the electrical diagram for the construction of the switchboard. In parallel, the programming of the PLC and HMI was performed. In the visual interface of the pre-pilot the user can control the peristaltic pumps, the liquid and gas solenoid valves and monitor the potentials and current of every MDC cell. Overall, the main characteristics of the control system-module are: IP66 Wall-mounting polyester enclosure; Safety equipment for devices and people protection; Power supply of the automation equipment, sensor and actuators; Power supply of the MDC; and Automation Equipment: PLC and HMI.

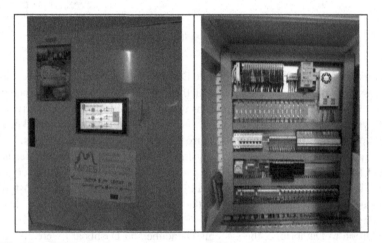

Fig. 11. MIDES pre-pilot control system switchboard: external and internal views.

Machine Learning using Databridge

As a sub-area of Artificial Intelligence (AI), Machine learning (ML) focuses on studying algorithms and statistical models used by computer systems to perform a specific task, relying on patterns and inferences. ML algorithms build a mathematical model based on sample data (training data), in order to make predictions or decisions without using explicit programmed instructions [22]. ML can be applied to various applications across business problems, together with Data Mining [23], focusing on exploratory data analysis for predictive analytics.

In MIDES, a Machine Learning (ML) approach with Statistical implementation was developed using the Databridge model to predict the Chemical Oxygen Demand (COD), by means of a software sensor (aka. soft sensor). The collected data as "training data" came from a total of 12 monitored variables, which included: conductivity, pH, redox, temperature, current generated, pump velocity, etc. (in catholyte, anolyte, saline tanks). Those parameters were measured every 1 min. Data from COD laboratory measurements were also considered, which have scarcely and infrequently measurements, of approximately twice a day. The COD soft sensor allows the estimation of COD in real time at every minute, allowing operators to take fast decisions.

To predict the COD, the variables parameters from saline, anolyte and catholyte tanks were used as input variables. Those were: the conductivity, pH, redox and temperature, for which a total of 16 samples were used to learn the COD predictive model. For such purpose, a partial least squares (PLS) model was used as the predictive model and deployed for operation in the Databridge module. The PLS was chosen as the predictive model, because it is robust to noise, correlated and irrelevant features, which are common issues in industrial applications.

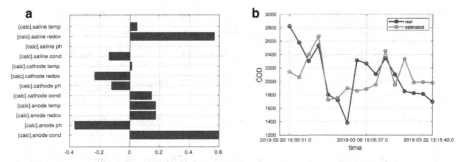

Fig. 12. (a) Coefficients of PLS model for COD prediction. (b) Predicted and real COD.

Figure 12 shows the coefficients of PLS model, and the prediction comparison, between real and estimated COD. Figure 12(a) shows the relevance of each input variable regarding the COD values. Figure 12(b) shows the coefficients of PLS model for COD prediction, it can be noticed that it has a pertinent and well-accepted agreement between prediction and real values, validating the ML approach.

6 The Operational Decision Making Support in the MIDES Pilot Desalination Plant

Operational Decision Support Systems have been extensively studied, including use-cases [24, 25], and show to offer decision makers accurate and appropriate conditions for better decisions, as they are provided with the needed information and results from simulated models of their analyzed processes, in order optimize operations as well as further implementation phases of respective systems and installation plants.

Within MIDES, the operational decision making support provided by the simulation analysis of the MDC and its process, as well as its automated operational control was of vital importance for the accuracy of its performance and data validation, as well as for the mitigation of risks in the pilot implementation stages.

The mathematical model of the MDC provided by the results of project work-package concerning *"Microbial Cell Design Engineering and Testing"* were further implemented in IPSEpro and integrated in the customized MIDES component-model library (MIDES_Lib), which was then used in the IPSEpro Process Simulation Environment PSE to create the comprehensive MIDES process model that was simulated, optimized and validated by all involved technical partners.

The concept of using IPSEpro as a Decision Support System (DSS) or coupled with a DSS for operational applications has its origin described in [15] and has been source of insights for further investigations in the DSS area (like in [20], for instance). Specifically, the analysis planned to support operations in MIDES was based on the insights and results of the respective simulated process model, as well as of its monitored control system. Figure 13 shows the considered steps used in MIDES for operational decision support, in order to allow pertinent actions to be taken to improve system performance envisaging more accuracy in the implementation phases of the project.

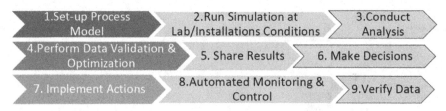

Fig. 13. Steps of the analysis planned for operational decision support in MIDES.

The overall simulated model implemented via IPSEpro, including its developed and deployed online platform, was then a useful asset for the project, in providing a decision-making tool concerning the integration of the MIDES process from Lab-scale to Pre-Pilot scale. The use of the simulated model of MIDES allowed the confirmation of expected parameters and values of the project implementation that then validated its implementation phase. Moreover, the results obtained in the analysis of the MIDES Pre-Pilot-scale MDC could be used for assessing the impact of up-scaling of changed component parameters in the next phase of the Pilot-scale MDC, and even analyzing the performance of the pilot plants. Likewise, the results of the MIDES Monitoring System allows project managers, end-users and decision-makers to have a clear picture of the operational situation of the pilot plants in order to be granted assistance on assertive potential decision-making options.

7 Summary and Conclusion

A cloud-based simulation platform for the IPSEpro process & simulation modelling environment has been implemented and its capabilities have been demonstrated. The experience gained with the system shows that it is well suited for realistic applications and can be particularly useful for collaborative project development, like MIDES.

In terms of the developed IPSEpro Online Platform, a major benefit of the cloud-based simulation platform is the fact that no software needs to be installed locally. This makes the system access easier, which is of particular benefit in collaborative research projects where a larger number of users needs to access the models. Additionally, there is no risk of version conflicts that can occur with locally installed software. Another advantage is that process models can be readily shared, and changes are available to all authorized users virtually instantaneously.

With the IPSEpro Online Platform and the Control Monitoring System implementation, we can assure that the shared MIDES process model for development collaboration among project partners is robust and accurate in performance and security aspects; and that the automatically monitored data presented to the end-users and pilot developers, including ML & statistical analysis, serve as important tools of decision-making with accurate levels of assertiveness. Those features enforce a rapid decision-making process on potential changes to be implemented in the model, impacting all stages of the project implementation from lab-scale, through pre-pilot to pilot stages.

Acknowledgment. The work presented has been partially developed within the EU Horizon 2020 research project MIDES (MIcrobial DESalination for low energy drinking water), under grant agreement N° 685793. The authors of this paper compose a sub-set of the full team in the MIDES project consortium and they wish to acknowledge the invaluable work done by all project partners.

References

1. Horizon 2020 Project MIDES https://midesh2020.eu/, MIDES - Microbial Desalination for Low Energy Drinking Water. DoA - Description of Action (H2020-NMP-2014–2015/H2020-NMP-2015-two-stage). In: European Commission - Horizon 2020 Project Grant Agreement No. 685793, pp 83231. Associated with document Ref. Ares (2016)1046927 (2016)
2. Thu, K., Chakraborty, A., Kim, Y., Myat, A., Saha, B.B., Ng, B.C.: AD Numerical simulation and performance investigation of an advanced adsorption desalination cycle. Desalination **308**, 209–218 (2013). https://doi.org/10.1016/j.desal.2012.04.021
3. Knowledge & Tools I 2030 Water Resources Group (http://www.2030wrg.org/knowledge-tools/). Charting Our Water Future. Online Report (2009)
4. Zamora, P., Ramírez-Moreno, M., Ortiz, J.M., et. al: Towards the world's largest microbial desalination cell for low energy drinking water production. In: Proceedings of the IDAWC19 International Desalination Association World Congress on Desalination and Water Reuse 2019/Dubai, UAE (2019)
5. Ramires-Moreno, M., Rodenas, P., et al.: Comparative performance of microbial desalination cells using air diffusion and liquid cathode reactions: study of the salt removal and desalination efficiency. Front. Energy Res. **7**, 135 (2019). https://doi.org/10.3389/fenrg.2019.00135
6. MIDES Consortium: Deliverable D3.1 Definition of the MDC mathematical model for the lab scale set-up operation. MIDES Project - Horizon 2020 Project Grant Agreement No. 685793 (2016)
7. MIDES Consortium: Deliverable D5.1 Comparison Study of Energy Performance for various Desalination Technologies. MIDES Project - Horizon 2020 Project Grant Agreement No. 685793 (2017)
8. MIDES Consortium: Deliverable D5.2 IPSEpro Model Library & Simulation of MIDES Process and Energy Consumption Study & Optimisation Analysis. MIDES - Horizon 2020 Project Grant Agreement No. 685793 (2018)

9. MIDES Consortium: Deliverable D5.3 – IPSEpro Models for MIDES Process & MIDES Web-based Process Model. MIDES Project - Horizon 2020 Project Grant Agreement No. 685793 (2018)
10. MIDES Consortium: Deliverable D5.4 – Overall MIDES automation project and Key Performance Evaluation. MIDES Project - Horizon 2020 Project Grant Agreement No. 685793 (2018)
11. Perz, E., Riesel, U., Schinagl H.A.: A new approach for modelling energy systems. In: ASME Cogen Turbo Power Conference, Vienna/Austria, 23-25 August (1995)
12. SimTech Simulation Technology: IPSEpro-PSE: Process Simulation Environment – User Manual. SimTech GmbH www.simtechnology.com
13. SimTech Simulation Technology: IPSEpro-MDK: Model Development Kit – User Manual. SimTech GmbH www.simtechnology.com
14. SimTech Simulation Technology: IPSEpro: Specification of the Desalination Library. www.simtechnology.com
15. Dargam, F.C.C., Perz, E.: A decision support system for power plant design. Eur. J. Oper. Res.-EJOR **109**, 310–320 (1998). Special Issue on Decision Support Systems
16. Aneke, M., Agnew, B., Underwood, C.: Performance analysis of the chena binary geothermal power plant. Appl. Thermal Eng. **31**(10), 1825–1832 (2011)
17. Karellas, S., Leontaritis, A.-D., Panousis, G., Bellos, E., Kakaras, E.: Energetic and exergetic analysis of waste heat recovery systems in the cement industry. In: Proceedings of ECOS 2012 – The 25th International Conference on Efficiency, Cost, Optimization, Simulation & Environmental Impact of Energy Systems, Italy (2012)
18. Perz, E., Erbes, M.: Process modelling of organic rankine cycles. In: ORC 2011, First International Seminar on ORC Power Systems, TU Delft, 22-23 September 2011 (2011)
19. W3C, HTML5: A vocabulary and associated APIs for HTML and XHTML. World Wide Web Consortium, W3C. https://www.w3.org/TR/2014/REC-html5-20141028/. (2014)
20. Kazim, B.A., Aydin, U.: A Multiple Criteria Energy Decision Support System, Technological and economic development of Econom (2011). ISSN 2029-4913 print/ISSN 2029-4921
21. Arevalo, J., Kenedy, M.D., Salinas-Rodriguez, S.G., Sandin, R., Rogalla, F., Monsalvo, V.M.: Pretreatment systems in desalination plants to reduce extreme events impact in drinking water production. In: Proceedings of the 1st pan-European Water and Sanitation Safety Planning and Extreme Weather Events Conference, Bilthoven, The Netherlands (2017)
22. Bishop, C.M.: Pattern Recognition and Machine Learning. Springer, Heidelberg (2006). ISBN 978-0-387-31073-2
23. Friedman, J.H.: Data mining and statistics: what's the connection? Comput. Sci. Stat. **29**(1), 3–9 (1998)
24. Nurminen, M., Suominen, P., Äyrämö, S., Kärkkäinen, T.: Use cases for operational decision support system. VTT Research Notes 2442, pp. 107-131 +app 5p (2008). Normative version: http://www.vtt.fi/inf/pdf/tiedotteet/2008/T2442.pdf
25. InteGRail FP6 Research Project: Intelligent Integration of Railway Systems (2005-2008). http://www.integrail.eu/documents/fs21.pdf

Overview

DSS, BI, and Data Analytics Research: Current State and Emerging Trends (2015–2019)

Sean Eom[✉] (iD)

Southeast Missouri State University, Cape Girardeau, MO 63701, USA
sbeom@semo.edu

Abstract. Over the past several decades, DSS has progressed toward becoming a solid academic field. Nevertheless, since the mid-1990s, the inability of DSS to fully satisfy a wide range of information needs of practitioners provided an impetus for a new breed of DSS called business intelligence systems (BIS). This paper examines the major differences among decision support systems (DSS), business intelligence (BI), and data analytics (DA). These three systems are different in terms of data, analytical environment, analytical tools, their focuses, and others. Next, the paper briefly describes the major characteristics of each. Third, A survey of DSS, BI, and DA is conducted covering the period of January 2016 through December 2019 to report publication trends of each system. The final section summarizes the findings and discusses their implications. This research provided a picture of what has happened over the past several years. The biggest accomplishment the DSS community achieved is that it provided the foundational concepts such as the DDM paradigm and the definitions of DSS. The DSS area has given the fruits of the DSS research to numerous other fields. Further, DSS and BI have a mild level of increased publications, while there is very sharp increase in DA (349%).

Keywords: Decision support systems · Business intelligence · Data analytics · BigData · Data warehouses

1 Introduction

The academic discipline of decision support systems (DSS) today has established itself as an academic filed of research and teaching, with contributions of more than 200 DSS researchers [1, 2], over the past 50 years. Until late 1990s, most textbooks had used "decision support systems" to cover the discipline. Nowadays, many of them have replaced "decision support systems" with "business intelligence" and/or "business analytics." For example, major information systems academic conferences such as the Americas Conference on Information Systems (AMCIS), European Conference on Information Systems (ECIS), and International Conferences on Information Systems (ICIS) create tracks on business analytics and business intelligence, in place of DSS. However, other DSS conferences use the term DSS, including the International Conference on Decision Support Systems Technology (ICDSST) and the International Federation of Information Processing working group (IFIP WG) 8.3 on Decision Support Systems.

© Springer Nature Switzerland AG 2020
J. M. Moreno-Jiménez et al. (Eds.): ICDSST 2020, LNBIP 384, pp. 167–179, 2020.
https://doi.org/10.1007/978-3-030-46224-6_13

Some believe that this is an indication that business intelligence systems (BIS) is a successor to DSS. We view DSS and BIS as two sides of the same coin and coexisting as two separate entities. BI is a more general system that has replaced specific DSS for data-driven problems where the intelligence phase of the decision is most important to the decision maker. BI and DSS have different origins, domains, impetus, and architectures. Academics always concentrated on "hard" problems, and this often means complex models, while vendors always want problems that occur widely, which is often data intensive. Over the past decades, with the changing technological environment of BigData, we have witnessed the emergence of the terms such as "BigData analytics."

The purposes of this paper are as follows. First, we examine the major differences among DSS, BI, and DA. These three systems are different in terms of data, analytical environment, analytical tools, their focuses, and others. Second, we briefly describe the major characteristics of each. Third, we present a survey of DSS, BI, and DA covering the period of January 2016 through December 2019 to report the publication trends of each system.

The next section presents the DSS Architecture, reference disciplines, and research sub-specialties. This is followed by the comparative analysis the major differences among DSS, BI, and DA. The final section summarizes the findings and discusses their implications.

2   Comparison of DSS, BI, and DA

The three aim to improve organizational decision making with different focuses, tools, and methods. The term "DSS" were invented by academic researchers to support decision makers to effectively deal with semi-structured decisions in the 1970s. On the other hand, BI and DA are jointly created by practitioners and software vendors to help organizations fulfill a wide range of information needs: information, decision making (structured, semi-structured, and unstructured), and knowledge creation (see Zaman [3], for introduction

Table 1. DSS, Business Intelligence, and (Big) Data Analytics

	DSS	BI	(Big) DA
Data	Database	Data warehouses	BigData
Analytical environment	Model Base User Interface	Data warehousing analytical environment	Advanced analytical techniques/Discovery analytics
Analytical tools	OR/MS	BI analytical tools (query, report, OLAP, Visualization & Data mining)	BI analytical tools & BigData processing tools (Hadoop, MapReduce, In-database analytics, In-memory databases Columnar data stores)

(continued)

Table 1. (*continued*)

	DSS	BI	(Big) DA
Focus	Semi-structured decisions	Information & Knowledge creation	Information & Knowledge creation
Origins	1970s	The mid-1990s	The early- 2000s
Who coined the terms?	Keen & Scott-Morton, Sprague & Carlson	Gartner, Inc.	Roger Mougalas from O'Reilly Media

Note: The analytical environment refers to the combination of hardware and software for building, accessing, and maintaining the analytical tools. The data warehousing analytical environment refers to the combination of hardware and software for accessing, and managing BI analytical tools.

to and evolution of BI). BI has been replaced by the new term "analytics" or "data science" [4]. In our view, DSS and BIS are independent information systems with distinct characteristics. Not all DSS implemented are created by BI tools. Many important DSS are the joint product of academicians and practitioners (Table 1).

2.1 DSS

The term, DSS, is created by academicians in business in 1970s. Over the past five decades, DSS has made progress toward becoming a solid academic field. Figure 1 is the summary of DSS research between 1969 and 2012 using author cocitation analysis (ACA), a subfield of informetrics [1]. The figure shows the data-dialogue-model (DDM) paradigm of Sprague and Carlson [5], organizational perspectives of DSS development [6], reference disciplines [7], and functional application development research [8–10].

2.2 BI

Another critical element that sets BI apart from DSS is data warehouses. BI systems consist of two major subsystems: the data warehousing environment and the analytical environment [4, 11].

The data warehouse is a central repository of reorganized, consolidated, and integrated data, which is subject-oriented, integrated, time-variant, and non-volatile, to give timely and reliable information and knowledge about the organization, its products/services, and its customers to help decision makers manage the organization effectively. The data warehouse is the architectural foundation of executive information systems (EIS), and business intelligence systems [12].

The data warehousing environment usually refers to the combination of hardware and software for building, accessing, and maintaining the data warehouse. The data warehouse environment extracts, cleans, transforms, transfers, and loads data from transaction processing systems. The management of the data warehouse consists of ensuring data quality, managing the systems (hardware, software, procedure, people, and data), and managing the metadata.

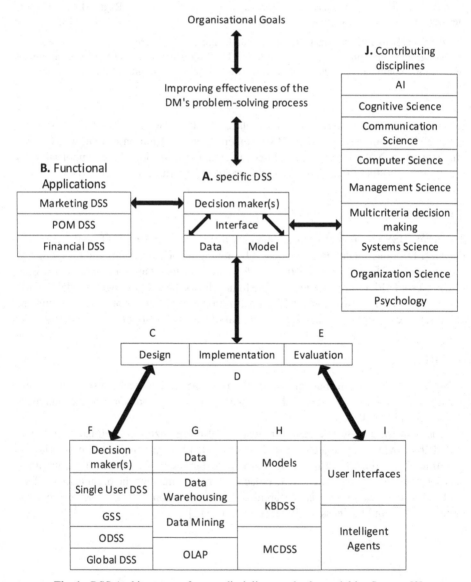

Fig. 1. DSS Architecture, reference disciplines, and sub-specialties Source: [2]

The analytical environment refers to the combination of hardware and software for building, accessing, and maintaining the analytical tools—query, report, online analytical processing (OLAP), visualization (e.g., pivot charts), data mining—in support of information extraction, decision making, and knowledge creation. The success of the data warehouse ultimately depends on the user's ability to carry out efficient and effective decision making in conjunction with several exploitation tools including the analytical environments.

2.3 DA

Due to the digital revolution, more and more data are easily and automatically captured and accumulated though the use of source data automation devices such as point-of-sale systems, optical barcode scanners, smart card readers, the Internet of Things (IoT), and social media. (Big) Data analytics (discovery analytics, or exploratory analytics) has two essential ingredients: big data sets and advanced analytic techniques.

Big data refers to a very large dataset whose size is too big to be processed by traditional hardware and software. Most big data are generated by machines in networks such as e-business networks (click stream data), social networks, the Internet of Things, many types of wireless sensor networks, and machine data (application logs, click stream data, web access logs, etc.). Bigdata may also include historic data from a data warehouse. No single computer hardware can store and process it with traditional database management software. Further, unlike traditional structured data that can be stored using a table of rows and columns, big data include unstructured data that are characterized the 4 Vs: Volume; Velocity; Variety and Veracity. These characteristics make it difficult to store, process, analyze the Bigdata using ordinary computing systems [4].

Advanced analytics is a collection of related techniques and tool types, usually including predictive analytics, data mining, statistical analysis, complex SQL, data visualization, artificial intelligence, natural language processing, and Bigdata processing technologies (such as MapReduce, in-database analytics, in-memory databases, columnar data stores).

In addition to (Big) data analytics, discovery analytics, or exploratory analytics, often large volume or large dataset analytics, and predictive analytics have been used interchangeably. With the increasing availability of bigdata in every industry, big data analytics have been applied to the following areas: smart city design, supply chain management, precision animal agriculture, genomic medicine, robotic industry, food and nutrition, healthcare research and also other type of interdisciplinary research, agile software development, etc. Our survey also reveals that many other terms are being used with specific focus on the nature of the dataset including engagement analytics, healthcare analytics, and learning analytics.

3 Emerging Trends

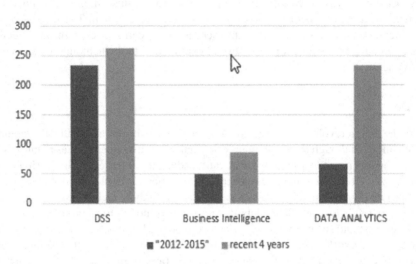

Fig. 2. Comparison of the number of Publications in DSS, BI, and DA

The first step is retrieving only "English" language, peer reviewed, academic journal articles published between 2012 and 2015 and between 2016 and 2019. To detect the trends, we divided into two periods. Prior research using the bibliometric approach covered up to 2012. This paper covers most recent 8 years. Figure 2 shows that research papers on DSS, BI, and DA have increased with different rates. DSS and BI have a mild level of increased publications, while there is very sharp increase in DA (349%).

3.1 DSS

Figure 2 shows that the number of published of DSS articles between 2016 and 2019 is 263. The majority (more than 82%) of them are being published in non-business fields such as engineering, animal biosciences, nursing, sports medicine, defense, child care, climatic changes, forensic science, etc.

The major accomplishments of DSS research are summarized in Fig. 1.

The emergence of DDM (Data-dialogue-model) paradigm (under A. specific DSS in Fig. 1): The DDM paradigm is concerned with the three major technology components required to build DSS – dialog management systems, data management systems, and model management systems [5].

Development of functional applications (under B. functional applications in Fig. 1) [8–10].

The emergence of the DSS framework from an organizational perspective (under C, D, E. in Fig. 1) [6].

The development of research subspecialties (under F, G, H, I in Fig. 1) [13–16].

Identification of contributing disciplines (under J in Fig. 1) [1].

During the period 2016-2019, here are summary of our bibliometric research findings published between 2016 and 2019.

DSS Developments in the Functional Area of Business

The first noticeable trend is specific DSS developments in the functional areas of business and management. This trend is due to the fact that all DSS reported here are OR/MS based systems. No other decision support technologies other than DSS are capable of developing specific DSS based on the DDM paradigm of Sprague and Carlson [5], where OR/MS models are indispensable components of DSS (Fig. 3).

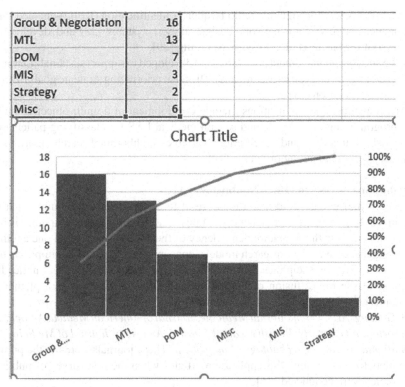

Group & Negotiation	16
MTL	13
POM	7
MIS	3
Strategy	2
Misc	6

Fig. 3. Distribution of DSS by functional management. MTL (Marketing, Transportation, and Logistics), POM (Production and Operations Management), Misc. (Miscellaneous Applications), MIS (Management Information Systems).

There were 47 applications in the corporate functional management area. Of the corporate functional management applications, group and negotiation support systems contains the largest number of applications (16), followed by MTL (13), POM (7), MIS (3), business strategy (2), and Miscellaneous applications (6). The MTL applications are further subdivided into several supply chain management support systems in the food industry[17], the offshore-wind industry [18], etc. In addition, MTL applications include social network information-based customer churn prediction DSS [19], and energy efficient ship operations support systems [20].

The group and negotiation support applications include the development of an automatic mechanism to support consensus reaching in the group decision making with heterogeneous preference structures [21], the development of a consensus-based multicriteria group decision model [22], and knowledge acquisition using group support systems [23].

The POM applications include a sustainable aggregate production planning model-based DSS for the chemical process industry [24], optimization model based DSS [25], collaborative web-based manufacturing operations support systems [26], and DSS for ranking the suitable contractors for a project by combining TOPSIS and AHP algorithms [27].

The MIS applications include radio frequency identification (RFID)-enabled three-echelon meat supply chain network design support systems [28] and developing a decision tool for business process crowdsourcing [29].

In the strategic management area, DSSs are developed to support sustainable strategic decision making in an electricity company[30] and in prioritizing strategic goals in higher education organizations [31].

Miscellaneous DSS applications include development of a multi-objective integer linear programming model [32], and several medical DSS for classifying patients into various risk groups [33], and developing a variable neighborhood search algorithm for the surgery tactical planning problems [34].

Contributions of DSS to Other Nonbusiness Area

Kaplan [35, p.4] wrote: "And it is a measure of its success in these inquiries that it is asked in turn to give its riches to other disciplines. ...... The autonomy of inquiry is in no way incompatible with the mature dependency of the several sciences on one another." DSS researchers have taken research models, concepts, theories, and assumptions from the various reference disciplines over the past several decades[7]. Eom [2] noticed the beginning to see the diffusion of DSS ideas and DSS applications in a plethora of diverse journals, including *Journal of Travel Research, Social Work Research, Simulation & Gaming, Archives of Internal Medicine, Journal of American Medical Association, Transportation Quarterly, British Medical Journal, American Journal of Medicine, and International Journal of Technology Management.* These journals were not the publication outlets for publishing DSS application articles when the first survey of published DSS articles were conducted [36].

This study shows that the majority (more than 82%) of DSS articles with "decision support systems" descriptor are being published in outside of business fields such as alcohol research, animal biosciences, biology, cancer education, child and adolescent psychiatry, child care, climatic changes, clinical therapeutics, crop management, defense, engineering, environmental policy, fish and fisheries, forensic science, forest research, lung cancer, mental health, nursing, physical geography, potato research, psychiatric services, sports medicine, telephone nursing, tropical medicine, etc. The DSS area is now giving the fruits of the DSS research to other fields.

3.2 BI

BI research is broadly concerned with the design, implementation, and use to achieve business goals. As such, BI research focused on identifying critical success factors [37], and justification [38] of BIS adoption and implementation as well as using BI for sensing opportunities for organizational innovation[39] and promote Business Unit Strategies, and domestic and international operations and business growth[40]. Data-driven DSS aims to produce BI that drives organizational innovations and business growth.

3.3 DA

To examine the development pattern and trends of DA research publications over the past 4 years, the survey results are summarized according to the area of application (Table 2). The application areas can be broadly divided into corporate functional management fields (23 applications articles) and other non-business areas (12 applications).

Corporate Functional Management Area
There were 23 BigData applications in the corporate functional management area. Of the corporate functional management applications, MTL contains the largest number of application articles published (9), followed by MIS (6), POM (5), and accounting and finance (2).

The MTL applications (9) are further subdivided into several subareas: supply chain management [41], customer relations management [42], and sustainable retail environment [43]. The MIS applications (6) are concerned with Big data-driven optimization for 5G mobile networks [44], IoT driven data analytics, and Big Data Analytics based on data from the multiple datacenters [45].

Data Analytics are applied in the POM area in building a better car company, designing analytics-based data products. BigData analytics are being used for smart production, and sustainable value creation [46]. Unstructured Data are being used to tidy up credit reporting and data analytics are applied to improve fraud prediction in accounting and finance. Big data analytics are improving dynamic capabilities and performance of firms (Table 2).

Non-Business Areas
There were 12 BigData applications in non-business areas. Of the non-business area, Education is creating and utilizing engagement analytics [47], which use online data collected from learning management systems' log data to predict learning outcomes and students' satisfaction. BigData analytics are being used to assess collaborative learning [48]. International applications research discussed data practices in South Africa [49] as well as challenges with big data analytics in service supply chains in the UAE [50]. In the area of city management, Amsterdam's Smart City Initiative was based on the utilization of BigData [51].

Table 2. Distribution of BigData analytics publications

Functional Management Areas		23
MTL	9	
MIS	6	
POM	5	
ACC/FIN	2	
STRATEGY	1	
Non-Business Areas		12
Education	3	
Int'l Application	2	
City Management	2	
Military	1	
Energy	1	
Healthcare	1	
Misc.	2	

4 Conclusions

Over the past several decades, DSS has progressed toward becoming a solid academic field. Nevertheless, since the mid-1990s, the inability of DSS to fully satisfy a wide range of information needs of practitioners provided an impetus for a new breed of DSS called business intelligence systems (BIS) as well as DA.

DSS research over the past several decades has evolved, and it has witnessed ongoing dynamic changes in the intellectual structure/disciplinary matrix of DSS research fields [1]. The academic discipline of DSS today has undergone numerous changes in technological and other environments. Before the era of big data, most data collected by business organizations could be easily managed by traditional relational database management systems with a serial processing system. Social networks, e-business networks, the Internet, and many other wireless sensor networks are generating huge volumes of data every day. The challenge of big data in both the business and political world has demanded a new decision support technology and new analytical environments. Accordingly, the academic discipline has responded to create a new field of BigData analytics.

This research provided a picture of what has happened over the past several years, based on the publication survey of DSS, BI and BigData Analytics along with previous bibliometric studies. We hope that results of this research provoke new thinking about the future direction of the DSS area. First, the biggest accomplishment the DSS community achieved is that it provided the foundational concepts such as the DDM paradigm and the definitions of DSS. The DSS area has given the fruits of the DSS research to numerous other non-business fields. Second, unlike a previous bibliometric study of DSS over the period (1969-2012), most DSS research publications has decreased substantially in many research subspecialties. Other than small number of application developments, the current survey indicated that there are little DSS research activities in the areas C through I in Fig. 1. The previous bibliometric study of DSS, over the period (1969-2012) concluded that some research subspecialties including most notably group support systems (GSS), model management, and individual differences/user interfaces, and implementation had disappeared or substantially weakened.

Past DSS research has contributed both breadth and depth to decision-making research. Future roles of DSS research will be even more important with the emergence of new decision and information technologies such as BigData, social and mobile computing and cloud computing to deliver more customer-centric and marketplace support. Data and Intelligence from BigData and data warehouses still need statistical and OR/MS modelling to create knowledge. We have seen many cases of datamining techniques applied to BigData and Data warehouse data. The future of DSS is to applying all statistical and OR/MS modeling to Bigdata and Business Intelligence data.

References

1. Eom, S.: Longitudinal author cocitation mapping: the changing structure of decision support systems research (1969–2012). Found. Trends® Inf. Syst. **1**(4), 277–384 (2016)
2. Eom, S.B.: The Development of Decision Support Systems Research: A Bibliometrical Approach. The Edwin Mellen Press, Lewiston, NY (2007)
3. Zaman, M.: Business Intelligence: Its Ins and Outs (2005)
4. Sharda, R., et al.: Business Intelligence and Analytics: Systems for Decision Support, 10th edn. Pearson, Boston (2015)
5. Sprague, R.H., Carlson, E.D.: Building Effective Decision Support Systems. Prentice Hall, Englewood Cliffs (1982)
6. Keen, P.G.W., Scott Morton, M.S.: Decision Support Systems: An Organizational Perspective. Addison-Wesley, Reading (1978)
7. Eom, S.B.: Reference disciplines of decision support systems. In: Burstein, F., Holsapple, C.W. (eds.) Handbook on Decision Support Systems 1: Basic Themes, pp. 141–159. Springer, Heiderberg (2008)
8. Eom, H.B., Lee, S.M.: A survey of decision support system applications (1971-April 1988). Interfaces **20**(3), 65–79 (1990)
9. Eom, S., Kim, E.: A survey of decision support system applications (1995-2001). J. Oper. Res. Soc. **57**(11), 1264–1278 (2006)
10. Eom, S.B., et al.: A survey of decision support system applications (1988-1994). J. Oper. Res. Soc. **49**(2), 109–120 (1998)
11. Eckerson, W.: Smart Comapnies in the 21st Century: The Secrets of Creating Successful Business Intelligence Solutions. The Data Warehousing Institute, Seattle (2003)
12. Eom, S.B.: Data warehousing. In: Zeleny, M. (ed.) The IEBM Handbook of Information Technology in Business, 1st edn, pp. 496–503, Thomson Learning, London (2000)
13. Eom, S.B.: Assessing the current state of intellectual relationships between the decision support systems area and academic disciplines. In: Kumar, K., DeGross, J.I. (eds.) Proceedings of The Eighteenth International Conference on Information Systems, pp. 167–182. International Conference on Information Systems, Atlanta (1997)
14. Eom, S.B.: The intellectual development and structure of decision support systems (19911995). Omega **26**(5), 639–658 (1998)

15. Eom, S.B.: Intellectual relationships between information systems and psychology. In: Proceedings of Associations for Information Systems 8th Americas Conference on Information Systems, Dallas, TX (2002)
16. Eom, S.B.: Intellectual relationships between the decision support systems area and cognitive science. In: The Eighth International Confernce of the Association of Information Systems SIG/DSS and International Society of Decision Support Systems: Trends in DSS Research and Practice, Porto Alegre, Brazil (2005)
17. Bocewicz, G., et al.: Traffic flow routing and scheduling in a food supply network. Ind. Manag. Data Syst. **117**(9), 1972–1994 (2017)
18. Mogre, R., et al.: A decision framework to mitigate supply chain risks: an application in the offshore-wind industry. IEEE Trans. Eng. Manag. **63**(3), 316–325 (2016)
19. Backiel, A., et al.: Predicting time-to-churn of prepaid mobile telephone customers using social network analysis. J. Oper. Res. Soc. **67**(9), 1135–1145 (2016)
20. Besikçi, E.B., et al.: An artificial neural network based decision support system for energy efficient ship operations. Comput. Oper. Res. **66**, 393–403 (2016)
21. Bowen, Z., et al.: Group decision making with heterogeneous preference structures: an automatic mechanism to support consensus reaching. Group Decis. Negot. **28**(3), 585–617 (2019)
22. Lima, A.S., et al.: A consensus-based multicriteria group decision model for information technology management committees. IEEE Trans. Eng. Manag. **65**(2), 276–292 (2018)
23. Pyrko, I., et al.: Knowledge acquisition using group support systems. Group Decis. Negot. **28**(2), 233–253 (2019)
24. Hahn, G.J., Brandenburg, M.: A sustainable aggregate production planning model for the chemical process industry. Comput. Oper. Res. **94**, 154–164 (2018)
25. Salam, M.A., Khan, S.A.: Simulation based decision support system for optimization. Ind. Manag. Data Syst. **116**(2), 236–254 (2016)
26. Hernández, J.E., et al.: A DSS-based framework for enhancing collaborative web-based operations management in manufacturing SME supply chains. Group Decis. Negot. **25**(6), 1237–1259 (2016)
27. Asgharizadeh, E., et al.: A combined approach of multiple attribute decision making for ranking and selection of project executive contractors in tenders. J. Econ. Manag. Perspect. **11**(3), 647–655 (2017)
28. Ahmed, M., et al.: A cost-effective decision-making algorithm for an RFID-enabled HMSC network design: A multi-objective approach. Ind. Manag. Data Syst. **117**(9), 1782–1799 (2017)
29. Thuan, N.H., et al.: A decision tool for business process crowdsourcing: ontology, design, and evaluation. Group Decis. Negot. **27**(2), 285–312 (2018)
30. Martins, C.L., et al.: An MCDM project portfolio web-based DSS for sustainable strategic decision making in an electricity company. Ind. Manag. Data Syst. **117**(7), 1362–1375 (2017)
31. Ivkovic, I., et al.: Prioritizing strategic goals in higher education organizations by using a SWOT-PROMETHEE/GAIA-GDSS Model. Group Decis. Negot. **26**(4), 829–846 (2017)
32. Guido, R., Conforti, D.: A hybrid genetic approach for solving an integrated multi-objective operating room planning and scheduling problem. Comput. Oper. Res. **876**, 270–280 (2017)
33. Nasir, M., et al.: A comparative data analytic approach to construct a risk trade-off for cardiac patients' re-admissions. Ind. Manag. Data Syst. **119**(1), 189–209 (2019)
34. Dellaert, N., et al.: A variable neighborhood search algorithm for the surgery tactical planning problem. Comput. Oper. Res. **84**, 216–226 (2017)
35. Kaplan, A.: The Conduct of Inquiry: Methodology for Behavioral Science. Transaction Publishers, New Brunswick, NJ (1998)
36. Eom, H.B., Lee, S.M.: Decision support systems applications research: a bibliography (1971-1988). Eur. J. Oper. Res. **46**(3), 333–342 (1990)

37. Yeoh, W., Popovic, A.: Extending the understanding of critical success factors for implementing business intelligence systems. J. Assoc. Inf. Sci. Technol. **67**(1), 134–144 (2016)
38. Popovič, A., et al.: Justifying business intelligence systems adoption in SMEs. Ind. Manag. Data Syst. **119**(1), 210–228 (2019)
39. Roberts, N., et al.: Using information systems to sense opportunities for innovation: integrating postadoptive use behaviors with the dynamic managerial capability perspective. J. Manag. Inf. Syst. **33**(1), 45–55 (2016)
40. Kuntonbutr, C., Kulken, M.: The effect of business intelligence on business unit strategies, international operations and business growth. J. Econ. Manag. Perspect. **11**(3), 1800–1807 (2017)
41. Kache, F., Seuring, S.: Challenges and opportunities of digital information at the intersection of Big Data Analytics and supply chain management. Int. J. Oper. Prod. Manag. **37**(1), 10–36 (2017)
42. Mahesar, H.A., et al.: Integrating customer relationship management with big data analytics in retail stores: a case of hyper-star and metro. J. Bus. Strat. **11**(2), 141–158 (2017)
43. Verma, N., Singh, J.: An intelligent approach to Big Data analytics for sustainable retail environment using Apriori-MapReduce framework. Ind. Manag. Data Syst. **117**(7), 1503–1520 (2017)
44. Zheng, K., et al.: Big data-driven optimization for mobile networks toward 5G. IEEE Netw. **30**(1), 44–51 (2016)
45. Eugster, P., et al.: Big data analytics beyond the single datacenter. Computer **50**(6), 60–68 (2017)
46. Dolan-Canning, R.: Sensor-based and cognitive assistance systems in industry 4.0: big data analytics, smart production, and sustainable value creation. Econ. Manag. Finan. Mark. **14**(3), 16–22 (2019)
47. Strang, K.D.: Beyond engagement analytics: which online mixed-data factors predict student learning outcomes? Educ. Inf. Technol. **22**(3), 917–937 (2017)
48. Williams, P.: Assessing collaborative learning: big data, analytics and university futures. Assess. Eval. Higher Educ. **42**(6), 978–989 (2017)
49. Winig, L.: A data-driven approach to customer relationships: a case study of nedbank's data practices in South Africa. MIT Sloan Manag. Rev. **58**(2), 3 (2017)
50. Khan, M.: Challenges with big data analytics in service supply chains in the UAE. Manag. Decis. **57**(8), 2124–2147 (2019)
51. Fitzgerald, M.: Data-driven city management: a close look at amsterdam's smart city initiative. MIT Sloan Manag. Rev. **57**(4) (2016)

Author Index